Scientific Manuscript and Report Writing:
A Pocket Guide for Getting Started

Adeline Boettcher, PhD

Copyright © 2021 Adeline Boettcher

All rights reserved.

ISBN: 9798735552833

CONTENTS

1 Overview   1

Part 1: Writing Organization Basics

2 Getting Started   5

3 Timelines and Keeping on Track   14

4 Identifying a Journal   24

5 Authorship   33

Part 2: Composing the Sections of Your Manuscript

6 Literature Search and Reference Management   39

7 Results   47

8 Introduction   58

9 Materials and Methods   70

10 Discussion   77

| 11 | Figures and Tables | 91 |
| 12 | Title and Abstract | 97 |

Part 3: Final Steps of the Writing Process

13	Self-editing: General Concepts	104
14	Self-editing: Language, Style, and Clarity	111
15	Writing a Letter to the Editor	121
16	Submitting Your Manuscript	129
17	The Editorial Process After Submission	133
18	Responding to Reviewers	141
19	Acceptance and Last Steps for Publishing	154
20	Handling Rejection	158
21	Other Resources	165

Scientific Manuscript and Report Writing:
A Pocket Guide for Getting Started

1
OVERVIEW

Writing a scientific document is both a challenging and fulfilling process. Whether you are in an academic or industrial setting, writing is a critical aspect to sharing, promoting, and engaging others in new scientific content. Scientific writing is an important skill that you can carry with you to other professions. Most scientific careers will require that you write manuscripts, technical reports, grants, or other documents. Thus, honing your writing skills is necessary for advancing in your career.

If you are reading this, you may be working on your very first report or manuscript, or you may have already published previously. Regardless of whether you consider yourself a novice or a much more experienced writer, there is content in this guide for you. As the chapters in this pocket guide are written, you can read them in any order, depending on what works best for your writing style or learning style.

You be in academia or working in an industrial setting. While publishing in journals may be less frequent for those of you working in industry, there will be times when you will need to put together a report for others in your department. Many of the principles in this guide for manuscript writing can be applied to report writing in an industrial setting.

Pocket guide overview

The chapters of this pocket guide are organized in such a way as to lead you through the process of writing a report or manuscript, from start to finish. While much of the terminology used throughout the guide is related to the composition of a scientific manuscript, the key concepts of this guide are applicable to a wide range of scientific writing projects outside of academia. This pocket guide is split into three sections: (1) writing organization basics, (2) composing the sections of your manuscript, and (3) the final steps of the writing process.

Part 1 provides details on the general topics you should think about as you are getting started with your writing. There is a lot of up-front preparatory work which consists of identifying a target journal, identifying your coauthors, and determining the best writing and editing strategies. Development of a "game plan" early on can help make the process of writing a manuscript more manageable.

Part 2 goes in depth into the specific details and workflow for composing the main sections of a manuscript. Each chapter in Part 2 is dedicated to a main manuscript section: abstract and title, introduction, materials and methods, results, and discussion, as well as a chapter for literature searches and reference management.

Part 3 includes details on the last stages of publishing. In this section, I provide tips on self-editing prior to submission. In addition, Part 3 contains information on the editorial process and how your manuscript is processed at a journal. This section also contains details on responding to reviewers, including what you will need to do when your manuscript is accepted or rejected.

Use this guide as a writing tool

In all, my goal in writing this pocket guide is to provide you with the tools you need to put together a great scientific report; with these same tools, you can continue to develop your writing skills. The themes within this pocket guide can be applied to a wide variety of scientific disciplines, as the presentation of scientific data uses the same principles across different fields.

The information presented within each of the chapters of this guide comprise a general framework for how each stage of the writing process can be approached. The concepts in this guide can serve as a starting point for immersing yourself in the writing process. My hope is that this guide will provide you with clarity and insight into the writing, editing, and publishing processes, helping you to write the best manuscript or report that you can.

2
GETTING STARTED

Identify your drivers and motivators

Sometimes getting started on writing can be the hardest part, but there are a few things that you can do to get the process started. Before you start writing, it is important to identify your organizational drivers and broader motivators. Organizational drivers are small and short-term organizational details that can help you feel in control of your writing projects. For example, my drivers include having a well-organized document with proper headers, completing to-do checklists, and having detailed notes from the literature I have read.

Broader motivators are the longer-term benefits of having a finished written piece. Some of my own motivators include learning new information, which I can do through performing literature searches, and sharing new information with others about the topic that I researched. Both of these things bring me satisfaction. By identifying my drivers and motivators, I can help keep myself on track during the writing process by reminding myself of my overall purpose.

In addition, I like to think of writing as an artistic process, with the finished manuscript being my piece of art. This mindset excites me, motivating me to write. Think about what aspects of writing you like; what helps to make the writing process more exciting for you? Try to identify some of your own drivers and motivators.

Here are some example motivators to consider:

- You will improve and expand your writing skills
- Publishing brings you closer to graduation
- Published manuscripts and reports boost your resume and CV
- You can share your work with others
- Performing literatures searches helps you learn more and can lead you to new research ideas

- You have the opportunity to teach others something new
- Having a manuscript with your name on it is impressive and gratifying
- Your written work can help advance projects within your lab

Create an outline and start a rough draft

Writing the first sentence of a report or manuscript can be one of the most challenging tasks of the entire writing process. It can be overwhelming to think about all the individual pieces of information that need to coherently come together. In addition, you may be worried about assembling this information in a timely manner. Thinking about either of those aspects can be particularly stressful when you are staring at a blank page.

One of the best strategies that I have come across is to just write a very rough draft "fact outline" to get the ideas out. To help set some organization in the beginning, start by labeling the four main sections: introduction, materials and methods, results, and discussion. These become the main points in your outline. If you are writing a literature review, think about four to six general topics you will plan to talk about in your review. You can always refine the review topics as you collect more information. Once you have these sections outlined, you can start adding in the facts that you have collected, which will eventually be incorporated into your document.

You may have difficulty focusing your thoughts as you being to write. To this end, it is a good idea to as yourself a few questions.

What experiments and data are you planning to present?

What methods will you need to include? Can you start to identify how you want to present the methods and associated results? Typically, each method associated with an experiment should have its own section and header.

What are the key points of your results? (See Chapter 7 for details on key points.) Each key point should have its own sub-header within the results section.

After you have asked yourself these initial questions, you can then start to jot down some of your main findings within these sections. Some journals allow sub-headers in the discussion as well. Typically, the discussion will be one of the last sections that you write because throughout the writing process, you will be thinking about concrete next steps and limitations of your study; you will not necessarily know this information until you have worked your way through the content of the document you are putting together. However, you can still add sub-headers to the discussion section during these beginning stages as you are brainstorming.

Develop a "fact outline" to build content

No matter what type of document you are writing, you will need to perform a review of the literature to provide a foundation for the content that you plan to present. You most likely already sifted through a lot of literature during the development, planning, and execution of your experiments. It is a good idea to create a written or typed physical list of publications that you have read. You will need to curate specific facts from these papers for your own manuscript or report. Start a file or folder on your computer to keep a log of all the publications that you read and collect information from.

One way to ease yourself into writing is to create a fact list. As mentioned above, find publications that are related to your own manuscript or report, and keep track of any citations within the publications that you read that look interesting or applicable. Write down methodologies or results from these other publications that are relevant to your study. As you collect this information, make sure to write the information in your own words.

Additionally, be sure to document the specific publication that the facts came from so that you can properly cite the work when it comes time to add your references (see Chapter 6 for more details on reference management).

There are numerous questions that you should ask yourself about the main sections of your paper as you are planning. These specific questions are outlined in the chapters for the introduction (Chapter 8), materials and methods (Chapter 9), results (Chapter 7), and discussion (Chapter 10).

In addition to finding specific facts that you can use for your project, you may also come across certain figures or table ideas that you like, or thought were presented in an informative way. Save or sketch those figures and tables as a foundation for continued idea development.

It is okay if your fact list is unorganized at the beginning of the process. When I am first starting a writing project, I will have random facts and citations all over a document. I organize these at a later point. It is important not to penalize yourself early in the writing process. Allow yourself to write without the anxiety of the content being disorganized. The important part at the beginning stages of writing is to just get the information on paper. Once you have some content written, it becomes much easier to string the ideas together into what will become your report or manuscript.

The initial draft does not need to be perfect. Without a doubt, there will be pieces of information that will be added, removed, and moved around within the document. For example, maybe you have a specific fact within the introduction section that ends up being moved to the discussion later; such changes are not only acceptable but are an important part of the thinking process when writing.

Anxiety in writing tends to come from the fact that the first draft will not be perfect, but no first drafts are perfect, not even those of seasoned writers. The processes of reading, writing, and editing are key to making a well-organized and well-written document; it does not happen all at once. You will begin to feel a sense of accomplishment once you start putting words on the paper.

It is also important to remember that the first draft is just that—the first draft. A document will go through many rounds of revisions in draft format. This revision process is necessary and helps ensure that you develop the best document that you can.

Now that we have seen some tactics for getting started on your writing project, we will talk about publishing timelines and other non-writing considerations. While this chapter lays a general foundation for how to get started, other chapters will have concrete examples of how to organize your sections.

Further reading:

- Kallestinova, E. D. How to write your first research paper. Yale J. Biol. Med. 84, 181-190 (2011).
- Gemayel, R. How to write a scientific paper. FEBS J. 283, 3882-3885 (2016).
- Setiati, S. & Harimurti, K. Writing for scientific medical manuscript: a guide for preparing manuscript submitted to biomedical journals. Acta Med. Indones. 39, 50-55 (2007).
- Meo, S. A. Anatomy and physiology of a scientific paper. Saudi J. Biol. Sci. 25, 1278-1283 (2018).

3
TIMELINES AND KEEPING ON TRACK

As scientists, sharing our work with the scientific community often gives us a rush of excitement. We have this new idea that we are eager to share with others; we want others to benefit from our research. However, the manuscript publication process takes time, and a long time may elapse between the initial submission and the acceptance decision. The time it takes to publish is widely variable and is dependent on many factors, such as the peer review process. In most cases, there are no hard and fast timelines for writing a manuscript, unless you are writing for a special collection, or your project has a due date. Some literature review articles can be written in as little as three months and published three months later, while original research articles could take two years from the initiation of writing to publication.

With such a variable timeline, there are a few important points to keep in mind as you write: *(a)* start and continue to write as you are performing your research and experiments, *(b)* focus on writing a good manuscript and do not rush yourself, *(c)* reach out to others for advice and guidance, and *(d)* remind yourself that composing a manuscript takes time, much like the research and experiments themselves.

I have been asked numerous times about how long it should take to write a manuscript. This is a difficult question to answer, though, because the length and depth of articles differ, as well as the fact that author writing processes can also vary. However, my general advice is that if you are in the midst of writing a manuscript, you should try to spend at least two hours writing, editing, or thinking about the work each day. Block it out on your calendar and dedicate this time to writing or planning.

Using checklists to manage your time

Depending on your writing style and process, you may find checklists to be highly beneficial. Some people love checklists; others hate them.

Personally, I feel a sense of accomplishment when I can check tasks off a to-do list. Checklists can also help with prioritizing tasks. My checklists allow me to see what needs to be done and I complete the essential and time-sensitive items in the day first. Then I can work on smaller, less urgent tasks in the later part of the day. I use this system for non-writing tasks as well, as it allows me to finish more stressful tasks earlier in the day.

It is important to assess how you feel on any given day and work on the task that will allow you to be most productive. For example, you should edit on days when you do not feel like writing and write on days when you feel inspired to do so. Editing often eases me into the process of writing, so I might do some editing first and then work on writing. As I edit, I can identify gaps within my writing, which allows me to consider what new content I need to add. Then, to find this information to add, I will read new relevant research, filling those information gaps and expanding my reference list at the same time. In this way, one task leads to another. So, even on days when I start out editing, I might end up actually writing. Whether you are writing, editing, or researching, it does not matter since all of these tasks contribute to the overall product and are all a part of the writing process.

Before you start writing, think of two to five objectives that you could complete in a two-to-four hour period. Write these on your checklist. Tasks could include writing about a specific procedure, making Figure 1, finding references for a specific topic, editing the introduction, or making a list of topics for the discussion section. These are the kinds of tasks that could be achieved in a single session of writing. Segmentation and completion of tasks will make the process of writing feel more manageable and listing such tasks on a checklist will help keep you focused.

If you have already collected all the data for a research article or if you are working on a review article, try to give yourself a goal end date for finishing the writing. Then, set smaller goals to break up the different parts of the manuscript; you can identify these different milestones on daily or weekly checklists. Such milestones could include completing tables, figures, a literature review on a specific topic, writing out the methods, or other similar tasks.

I have known individuals that have "deadline anxiety" and do not function well if they have a deadline. If setting a deadline is not the best strategy for you, try tackling tasks one by one using a checklist in a way that works best for you. The important part is that you put time and effort into the project each day.

Getting in the zone and minimizing distractions

When I edit, it can take me thirty minutes to an hour to really get in "the zone" where I can critically assess what I am reading. If you do not allocate enough time to focus at the beginning of a work session, you may reach the end of your work period feeling like you have not made enough progress. To that end, I would suggest that when you sit down to write or edit, you should do it for at least one hour.

Everyone is distracted by different things, and this can affect your work environment. Depending on what things distract you, try working at a library, coffee shop, or at home. If music helps you focus, listen to music. If you know that social media distracts you, put your phone in a different room and close all your social media tabs. You can also wear headphones or earbuds simply so other people do not talk to you. Identifying things that you know can be a distraction and try to minimize them.

As mentioned in the section above, start with a checklist and identify the task that you feel most excited about (or is the easiest task!); doing so may help you more easily get into "the zone".

Additionally, it is important to take regular breaks to reset and refresh. If you find yourself becoming distracted—looking at irrelevant webpages or looking at your phone—this is often a sign that you need to take a break. On the other hand, sometimes when you are deep in the zone, you will be able to work for longer stretches without breaks. Some options for regular breaks may include a 10-minute break every hour or a two-minute break after 20 minutes. Find a break schedule that works well for you.

Check in with your project leaders and other collaborators regularly

Your advisor, project leader, or principal investigator (PI) and other coauthors can be helpful sources of information and should be able to give you a second (or third) opinion on writing style or data presentation. Additionally, your coauthors and advisors should have some experience with choosing journals as well, so make sure to reach out to them and keep them updated as you continue working through your manuscript.

There are three main check-in points you should have with your coauthors:

1. When you are first planning the outline of your manuscript
2. When you have written an initial draft
3. When you have a final draft ready for submission

During the planning and drafting phases of writing, I will often sketch out what I envision certain figures and tables will look like in the final document. Then I will show my coauthors my ideas *before* I commit a lot of time to working on them. Often coauthors will provide you with new ideas or suggestions that may be slightly different than your original ideas. Additional information about coauthors can be found in Chapter 5.

Once a manuscript is complete and near being ready for submission, you will need to send the draft out to all coauthors. Because your coauthors will likely have slightly different areas of expertise, they can help you in identifying gaps that may exist in the manuscript before you submit it. For example, during my graduate work, my project involved the development of swine biomedical models for different areas of disease research. My research team consisted of experts in veterinary medicine, immunology, genetics, animal science, and molecular biology. It is important to realize that you, yourself, do not need to be an expert in everything. Multidisciplinary teams exist to allow research groups to have a wide breadth of ideas and contributions. Reaching out to experts in different fields can help give you a wide range of quality feedback on your writing.

Generally, when you send out a draft to others,

it is a good idea to give your coauthors one to two weeks to respond so as to allow them enough time to read the work and provide feedback. While you may feel overwhelmed by the edits that you receive from everyone, remember that the edits will improve your document. Also note that you should not feel discouraged if you receive a lot of edits. Remember that you are receiving comments from a variety of experts with different scientific backgrounds. The point of asking for feedback is so that you can improve the work. In the end, implementation of this feedback will improve the writing and messaging in the manuscript or report that you are putting together.

Overview of editorial timelines

Just as the time it takes to complete a manuscript varies, there can also be variations in the amount of time it takes for your manuscript to undergo the peer review process after you have submitted it. In an ideal world, a newly submitted manuscript would undergo peer review the day it is submitted, and we (the authors) would get a decision within a few days. However, we all know the process does not work that way. Quality reviews do not always happen in a rapid manner.

Chapter 17 will provide an in-depth look at what happens during the editorial process after you have submitted your manuscript, but the next few paragraphs will provide a high-level overview of this process as well as a rationale for the certain timelines you may experience when awaiting a decision on your manuscript.

In most cases, within the first two weeks after you have submitted a paper, the manuscript will undergo an internal review process and a quality check. This quality check includes ensuring correct word counts, confirming that the correct number of figures and tables were submitted, verifying that all coauthor information was submitted, and performing other similar inspections. If all is approved at the quality check, the manuscript will then go to the editorial board for an initial discussion.

When the editorial board receives the manuscript, they will assess the scientific merit of the research within the study. If the manuscript is determined as having merit and is relevant to the journal, then potential peer reviewers will be contacted. Assuming the first set of selected reviewers agree, the peer review process may take somewhere in the range of two to three weeks.

After the peer reviews are submitted, the editorial board reconvenes and discusses their decision of your manuscript. If all goes smoothly through each of these steps, you may expect to hear a decision some seven to nine weeks after you have submitted your manuscript. However, all of the people involved in the process (publications staff, editorial staff, peer reviewers) are just that—people. Holidays, illnesses, vacations, business projects, and other life events could cause small delays at any point in the processing of your manuscript. For this reason, initial manuscript decisions can sometimes take up to three or more months.

Redundant saving

Although saving your work is not directly related to the timeline of your project, not saving your work can certainly affect your timeline, as well as your motivation. There is nothing is more heartbreaking than losing your work. Losing your work can affect your overall motivation as well as your "flow" on the project. To that end is important to continually save your documents. Save your content in three places:

1. Locally on your computer
2. In a cloud application or email
3. On a USB or external hard drive

You may even wish to have more than one USB drive—one that you keep at work and one that you leave at home. You will thank yourself for saving the document in multiple places if your computer ever crashes or if any of your files become corrupt. I always save locally and then email myself the documents before I close my computer for the day.

4
IDENTIFYING A JOURNAL

Overview

Choosing a journal is one of the more fun parts of the writing process. There are a variety of factors that you may weigh when deciding which journal to submit to, including:

- journal scope,
- impact factors,
- cost to publish, and
- peer review time.

You can talk to your project leader or other collaborators about which journals they have published with in the past. You can additionally ask them about their experiences with each of those journals as well. Experts in your discipline will likely have a good understanding of the different journals within your field.

Journal scope

There are a few easy ways to identify journals that match the scope of your research. Think about some of the articles that you read while developing the ideas for your project—what types of journals were they published in? Would any of those journals be good candidates for your work? Also look at the citations within some of your favorite papers—some of those journals could be good submission candidates as well. You can also do a simple search online to identify possible journals. There are many websites that provide lists of journals for different subject areas.

Once you identify a few potential journals, look on their websites to find out if they have published papers within your general subject area. Use some of the search terms that you expect to have within your title and abstract (see Chapter 12) on the journal website to see what types of studies appear from the search. If you see that there are previous publications that are similar to the study you are writing about, then there is a good chance that your research topic would be a good fit for the journal.

Impact factors

An impact factor is a measure of how often articles from a journal are cited in other articles. An impact factor is calculated based upon the number of articles a journal has published within a two-year period and how many times those articles are cited within the following year. For example, if Journal A published 50 journal articles (citable items) in 2017 and 2018, and in 2019 there were a total of 346 citations from manuscripts in Journal A (published during 2017-2018) in other manuscripts (from other journals), the impact factor for Journal A would be 6.92 (346/50).

Typically, researchers submit to high impact journals first, and then resubmit to journals with lower impact factors if rejected. In the sciences, the two journals that come to mind as having high impact factors are *Science* and *Nature*; however, both journals are multidisciplinary (i.e., their topics range from botany to astronomy to biomedical engineering), which plays a role in their impact factor. There are also discipline-specific high impact journals like *Radiology*, *Journal of the American Chemical Society*, and *Advanced Materials*.

As you are deciding which journal to submit to, it is important to critically assess your work. If your manuscript ends up being rejected, consider the fact that you may need to reformat it before submitting to another journal. Reformatting a manuscript can take a lot of valuable time.

Formatting and reformatting can be stressful and, in some cases, an unnecessary process. Before you submit your paper, create a plan with your project PI for a journal "succession line". This succession line should consist of two to four other journals you could submit to in case your manuscript is rejected from your first-choice journal. Having this list allows you to be prepared to resubmit elsewhere. Many manuscripts are not accepted to the first journal they are submitted to, so do not be too disappointed if you receive a rejection. The very first paper I published as a graduate student was rejected from two journals before being accepted; it was a process that took over a year and a half! Chapter 20 contains more details about rejection decisions.

It is important to remember that if your manuscript does not get accepted to a high impact journal, it does not mean that your manuscript does not have value or that it will not be accepted elsewhere. The review process for all journals is rigorous and journal editors need to ensure that anything they consider for publication fits within the subject area of their journal (see Chapter 17).

Submission and publishing costs

There are several different types of fees that you may come across in the publishing process. These generally fall into three main categories:

- submission fees,
- article processing fees, and
- fees for colored images and supplemental material.

Journal submission fees are not universal for journals, but some journals may charge in the range of $50 to $500, and these fees that are non-refundable if a manuscript is rejected. Many journals do not have submission fees, but it is best to assess ahead of time if your journal of choice requires payment for submission or not.

Article processing fees are fees which pay for the editing, production, and online presentation of an article. If you are submitting to a journal that is solely subscription-based and are not opting to publish an open access article, you likely will not have to pay an article processing fee. Journals typically generate revenue through either subscription fees or through article processing fees for open access articles.

Article processing fees are required for open access articles. At the time of this writing, article processing fees for open access articles can range anywhere from $1,200 up to $11,390 for top-tier journals. As you are preparing to submit to a journal, you should decide if you want to publish open access and, if you do, how much the article processing fee will be.

Journals may charge extra for colored figures or supplemental material. If a journal still prints physical copies of each issue, they may charge for colored figures. Additionally, since supplemental information requires that a journal store more information for an individual article, the journal may also charge for supplemental material. These fees may be a blanket fee for any supplemental material, or the fee may be charged based on the number of extra figures or tables that you have.

Peer review times

You may have heard that there are certain journals that have long peer review times. I have even had past collaborators suggest that I not submit to specific journals because of prior experiences they had with the peer review process. For this reason, it is a good idea to talk with others in your research group about their past publishing experiences to get an idea about the decision timelines of different journals since this might also affect where you submit to.

A journal's decision timeline may also influence your decision of where to submit to. There are a few ways that you can find general decision timelines for journals. Some journals show their statistics for average decision times directly on their website. Additionally, published articles will typically have specific dates listed on the publications themselves detailing when the authors initially submitted the article, when the first decision was made, and when the final decision was made (after peer review and revision). You might choose a journal with a longer decision timeline because the journal seems to otherwise be a very good fit for your field of study. The decision timeline is another factor to consider when deciding where to submit your work.

Journal-specific formatting

One other factor you should consider as you are selecting your journal is the journal-specific formatting. Even if the different formatting requirements do not impact your decision of where you submit, they are important requirements to be aware of.

You do not necessarily need to have a specific journal picked out before you start writing. In fact, it is likely that you will change your mind about which journal you plan to submit to as you go through the writing process. However, it is still important to look at the guidelines for submission before you start writing. There are often word limits that can be difficult to attain once you have a full written manuscript.

Below is a list of some journal-specific formatting that you may come across:

- Word limits for abstract
- Structure of abstract
- Word limits for main text
- Citation style
- Limit on number of tables
- Limit on number of figures
- Structure of main text

Differences in word limits are probably the most common variance that you will see between journals. Make sure to abide by the guidelines. There have been times when I have needed to cut 1000 words before submitting to a new journal. If you can determine the guidelines early, it makes it much easier to prepare your submission for that specific journal. Journals may send a manuscript back to you to edit if your word counts are too high. If this happens to you, make sure you save all of the different versions of the text. If you edit to reduce the word count, save all the removed text in a separate document in case you need it again later. You can use these separate documents as a resource for future writing or editing.

Further Reading:

- Masten, Y. & Ashcraft, A. Due diligence in the open-access explosion era: choosing a reputable journal for publication. FEMS Microbiol. Lett. 364, (2017).
- Bahadoran, Z., Mirmiran, P., Kashfi, K. & Ghasemi, A. Scientific Publishing in Biomedicine: How to Choose a Journal? Int. J. Endocrinol. Metab. 19, e108417 (2021).

5
AUTHORSHIP

Different roles in research

Throughout the planning and execution of your projects you likely spoke and worked with quite a few individuals. When it comes time to compose your manuscript, you will need to be able to identify which of these individuals will be coauthors on your manuscript. Below are definitions of collaborators, coauthors, and contributors and how they relate to your projects:

Coauthor: An individual who has contributed enough to the work to take public responsibility for the content. These individuals are included as part of the author list. Coauthors may be involved in the writing of the manuscript.

Collaborator: An individual from the research group who contributed significantly to the work that is not an author. These individuals are named within the acknowledgement section.

Contributor: An individual that participated in any part of the research work. Contributors consists of both collaborators and coauthors.

While some of these definitions overlap slightly, it is important to be aware of the distinctions between these different people.

Throughout the course of your project, you will likely work with numerous discipline-specific experts as your contributors. As mentioned previously (Chapter 5), a project team consists of diverse members that can bring different perspectives to a project. It is best to have many discipline-specific experts on your team because these individuals can substantially enhance the quality of the research that is produced.

The next page shows a list of criteria that can be used to determine if an individual should be considered as a coauthor of a manuscript.

Did these individuals…

1. Conceptualize or help design the project?
2. Perform data acquisition, analysis, or interpretation?
3. Draft sections of the manuscript?
4. Critically revise or contribute intellectual content?
5. Perform statistical analysis?
6. Obtain the funding for the project and provide oversight?
7. Supervise and lead the project?
8. Perform administrative, technical, or material support?

If there were other individuals who contributed to your project, but do not fall into any of the above criteria, you can mention them in your acknowledgments section.

Co-first authorship

In recent years there have been many strides made towards more collaborative research projects. As such, there are now an increased number of publications with more than one first author. More journals are adopting the practice of allowing two first authors, as well as two corresponding authors (i.e., the PI(s) of the project). Determining who of the two individuals should be listed first can be a difficult decision. It is best to approach this decision in an objective manner, assessing the work contributions from both individuals. Some points to think about if you are in this situation are the following:

- Assess the novelty of the contributions from both individuals
- Determine the time each individual spent on the project
- Consider who coordinated and executed a majority of the subprojects
- Assess who wrote a majority of the manuscript

Of course, all these points are on a moving scale, but you will want to keep them in mind as you plan your author list. If you are involved in a study where co-first authorship may be warranted, it is best to discuss this prospect with the project PI.

Determining coauthor order

As you are working through your manuscript, it is important to continually assess which coauthors you will need to include, and in which order they will be presented in the author list. The individual who coordinated, planned, and executed most of the experiments should be the first author on a manuscript. The individual who oversaw, mentored, and provided the funding for the project is typically listed last as the corresponding author. Everyone listed in between are those contributors who provided substantial data and guidance throughout the research project.

Determining the order of the authors can be difficult, as you want to ensure that all individuals are treated fairly in the process. Individuals who performed experiments and contributed data to the project should be listed immediately after the first author. To do this, you will have to assess which of your contributors provided the most data, time, and interpretation for the study. Others who provided general guidance, help, and coordination in the project may be listed towards the end of the list.

Journals also typically ask that you provide a list of author contributions. Writing out these contributions beforehand may help with determining the order of the author list. As you make your list of coauthors and roles, consider which experiments and contributors played key roles in bringing the study results together.

Further Reading:

- Dulhunty, J. M., Boots, R. J., Paratz, J. D. & Lipman, J. Determining authorship in multicenter trials: a systematic review. Acta Anaesthesiol. Scand. 55, 1037-1043 (2011).
- Aliukonis, V., Poškutė, M. & Gefenas, E. Perish or Publish Dilemma: Challenges to Responsible Authorship. Medicina (Kaunas). 56, 123 (2020).
- Christiansen S, Iverson C, Flanagin A, et al. AMA Manual of Style: A Guide for Authors and Editors. 11th ed. Oxford University Press; 2020.

6
LITERATURE SEARCH AND REFERENCE MANAGEMENT

Literature searches are essential for building the content of your manuscript or report, and managing your references appropriately during the early stages of the literature review can make your work much easier when it comes time to cite studies within your manuscript. There are three main phases of performing a literature search:

1. Identify questions that you will need to address in your manuscript.

> During this phase, you will need to brainstorm about what potential questions a reader will have about your study. Identifying these

questions will guide you in gathering appropriate information so that your reader can understand and appreciate your work.

2. Determine where the gaps are in the literature.

> These literature gaps are the driving force behind your work. Your research group likely already had these gaps identified before you started your research, but it is important that you explain these gaps within your manuscript to provide context to your reader about why you performed your study.

3. Understand where your research will fit within the existing literature.

> Lastly, you will need to explain the broader impact of your work to your readers. You can do this by gathering information about similar studies that have been performed and showing how your work is a stepping-stone for the next phase of research.

Anticipate your readers' questions

There are multiple ways to perform a literature search. The main goal is to collect resources that you can use to describe your research in such a way that your readers can understand and appreciate the findings. Try to identify what points you will need to walk your reader through so that they can understand the topics you are going to talk about.

Below are some sample questions to ask yourself:

- Why is my new area of research important?
- Am I describing a new methodology? What might my readers need to know about my methodology?
- How can I describe my model system to my readers?
- Why is the animal model I used a good fit for my research?
- Am I describing some type of drug? What should I tell my reader about the drug? What is the mechanism of action for the drug?
- Am I going to need to describe a cellular mechanism? What information will be reader need to know to understand the mechanism?
- What are the impacts of the results from my study?

The questions you ask yourself will vary depending on your field. However, all of the questions fall within three main categories:

1. What the research is about
2. How the research was performed
3. Why the research is important

Break apart your research and figure out what questions a reader may have about your subject area. Search for publications that help you to answer these questions. The chapters about assembling the introduction (Chapter 8) and discussion (Chapter 10) provide additional questions that you should think about as you are writing these sections.

Reference organization

Organizing references is an important task—especially when you are trying to remember the author's name and the topic of the manuscript. I have found that having a notebook or word-processing document with bullet points from different papers, along with the author's name and publication year, is a helpful resource. When you are writing these initial notes, they do not necessarily have to be organized. Once you collect a few different facts from different publications, you can start to organize them based on subject content. The most important aspect in the beginning stages of the literature review is to accurately document the original publication that the information came from.

In the past, I have put notes from the papers I read directly into the same document in which I am writing my manuscript. I typically have two to three important bullet points that I pull from each publication. You could even start a very general outline of your manuscript with these collected bullet points. Write out the main sections (introduction, materials and methods, results, discussion) and, under each section, make a list of references and bullet points that you will want to expand on once you start writing. This way you do not have an additional step to take once you are ready to start writing. Chapters 2 and 3 have additional information on the initial phases of writing and organization.

Find a citation manager

In the first paper I published, *I manually typed out all of my citations*; between formatting and double-checking author names, I wasted a lot of time. Manually typing citations can also introduce a lot of errors within author names, the publication title, and other parts of the citation. Do not subject yourself to this workload. Instead, use a citation manager.

A citation manager is a piece of software to which you can upload citation files (commonly RIS or NBIB files) that contain the reference title, authors, journal, issue, volume, year, and other relevant information. These file types can be downloaded from journal websites directly or they can be downloaded from the article entry on PubMed, Google Scholar, Scopus, ScienceDirect, or whichever search engine you use. There are usually plug-ins from different citation mangers that can be added to the word-processor program that you are using. Adding in a reference is usually as easy as typing in the first author's last name and adding the citation within the text. Another nice aspect about citation managers is that they are programmed to format your citations based upon specific journals. There is usually a menu where you can choose your journal of interest and the program will reformat in-text citations as well as the references list.

There are a lot of different citation managers available. The three that I have commonly heard other researchers use include EndNote, Mendeley, and Zotero. Explore around online to find the citation manager that works best for you.

A summarized checklist

There is no right or wrong way to perform a literature search. However, for a structured method, follow the guidelines below:

- Identify and create a list of questions your audience may have about your research
- Categorize the questions by subject matter
- Determine in which areas of the manuscript those questions may arise (introduction, results, etc.)

- Write down bullet-pointed facts from your references under each question
- Find and use a citation manager

Further Reading:

- Ackerman, M. B. Why perform a literature review? American journal of orthodontics and dentofacial orthopedics : official publication of the American Association of Orthodontists, its constituent societies, and the American Board of Orthodontics vol. 145 549 (2014).
- Syam, A. F. Tips and tricks in writing review article. Acta Med. Indones. 39, 143-144 (2007).
- Younas, A. & Ali, P. Five tips for developing useful literature summary tables for writing review articles. Evid. Based Nurs. 24, 32 LP - 34 (2021).

7
RESULTS

Manuscript section overview

Next, we are going to shift gears to discussing topics related to writing the body of your manuscript. Having an organized plan from the beginning will make the writing process easier. It can be easy to get bogged down at the start because there are so many pieces of information that need to be brought together. The next few chapters will discuss the process of assembling each of the major sections of a manuscript and address what key questions and points should be made in each. With the exception of this current chapter (results), the next chapters are presented in the order that you will present your information within a manuscript: introduction (Chapter 8), materials and methods (Chapter 9), and discussion (Chapter 10). This is the order in which I have typically approached writing a research paper, as well; however, there is no prescribed order for writing a manuscript. I plan out and write my results section first (and therefore have placed this chapter about results first) because identifying the content that you will present in the results section can help lay the foundation for the content of the other three sections.

Preparing your results content

The process of performing experiments, compiling data, analyzing data, and forming cohesive results for a single project may take years. As this information piles up, some form of organization is required to help bring the individual components together and develop the story of your research.

One approach that you can take as you begin to write your manuscript is to identify the main two to six key results that you plan to showcase in your paper. While you may have run over 20 experiments, you need to be able to focus on the main overall result from all of these experiments. If someone were to give you 30 seconds to summarize your key findings, what would you tell them? These key points can be used in both the section headers of your results section and the figure titles (with some rewording).

To get started, try to write one paragraph (three to four sentences) for each key point of your manuscript. These do not have to be perfectly written paragraphs. Just get the words down first; they can be edited later. Take below for example:

In the first part of our project, we wanted to mutate the IL2RG gene in our existing ARTEMIS$^{-/-}$ pig line. This is because an ARTEMIS$^{-/-}$ IL2RG$^{-/Y}$ pig would have a T$^-$ B$^-$ NK$^-$ cellular phenotype, which we are interested in for further developing biomedical models. We were able to develop these pigs using site-directed mutagenesis of IL2RG with CRISPR/Cas9.

This paragraph can then be shortened to:

Generation of ART$^{-/-}$ IL2RG$^{-/Y}$ SCID pigs by site-directed CRISPR/Cas9 mutagenesis of ART$^{-/-}$ fetal fibroblasts

The shortened phrasing above can then be used as a section header within the results section. As you write your key points and headers within your results section, make sure to be specific. Make it easy for your readers to know what you did and what you plan to discuss within your results sections. Remember that readers may quickly skim through your paper without reading all the details first. As headers are typically bolded in the final publication, having these key points pulled out helps readers take away the overall point of the paper.

Identifying the main points that you plan to present in the results section also provides you with a clear path for gathering information for the rest of the paper. Importantly, these points help to lay the foundation for the specific information you will need to describe in your introduction in order for your audience to understand and appreciate what you are talking about in the results section.

To guide you further, look through some of the papers that you have read recently. Read the headers in the results sections. Are they descriptive? You may come across some headers that are vague. However, descriptive headers make it quite easy for your reader to know exactly what they will be reading about within each of the results sections. Additionally, once you have these key points written out, you can use them to start developing the outline for your manuscript.

The next few pages show some examples of results headers.

Example set 1:

- *Bioinformatics analysis of CCNB1 expression in HCC tissues and normal liver tissue*
- *CCNB1 knockdown promotes apoptosis and suppresses proliferation in HCC cell line HepG2 and SMMC-7721*
- *CCNB1 knockdown inhibits HCC cell migration and invasion*
- *MiR-144 inhibits CCNB1 expression by binding to CCNB1 3'-UTR*
- *MiR-144 mimics elicit similar proliferation repressive effect in HCC cell*
- *MiR-144 overexpression suppresses the tumor formation of SMMC-7721 cells in nude mice*

Text from: *Gu J, Liu X, and He Y. (2019). MicroRNA-144 inhibits cell proliferation, migration and invasion in human hepatocellular carcinoma by targeting CCNB1. Cancer Cell International. 19: 15. Text from original article. Reprinted under a Creative Commons Attribution 4.0 International License.*

Example set 2:

- *Significant Elevation of Serum Hepcidin in DS Subjects*
- *Iron Accumulation Is Visible in DS Senile Plaques*
- *Hepcidin Protein Accumulates Around the Senile Plaques in AD and DS Brains*
- *Hepcidin and Ferritin Expression in Astrocytes Close to the Blood Vessels*

- *Selective Population of Microglia and Meningeal Macrophages Are Involved in Brain Iron Homeostasis in DS*
- *IL-6 Could be Involved in Host Defense in AD and DS-Brain*

Text from: *Raha AA, Ghaffari SD, Henderson J, et al. (2021). Hepcidin Increases Cytokines in Alzheimer's Disease and Down's Syndrome Dementia: Implication of Impaired Iron Homeostasis in Neuroinflammation. Frontiers in Aging Neuroscience. 13: 653591. Text from original article. Reprinted under a Creative Commons Attribution License.*

Example set 3:

- *Genome assembly and annotation supported by comparative and functional genomics*
- *Limited evidence for lateral gene transfer and no evidence of endosymbionts in the Stomoxys genome of a laboratory-colonized strain*
- *The Stomoxys immune system encodes gene family expansions that may reflect adaptation to larval development in microbe-rich substrates*
- *Gustatory and ionotropic receptor gene family expansions support importance of bitter taste perception in Stomoxys*
- *Expansion of the long wavelength-sensitive Rh1 opsin subfamily in Stomoxys with evidence of tuning for diversified wavelength sensitivities via substitution*

Text from: *Olafson PU, Aksoy S, Attardo GM, et al (2021). The genome of the stable fly, Stomoxys calcitrans, reveals potential mechanisms underlying reproduction, host interactions, and novel targets for pest control. BMC Biology. 19: 41. Text from original article. Reprinted under a Creative Commons Attribution 4.0 International License.*

Introduction and conclusion statements

Two important components to add to your results paragraphs (if space allows) are introductory and concluding statements. Try not to start paragraphs by saying "Figure 1 shows X." Rather, start with "We were interested in X and, thus, did Y. We found A, and results are reported in Figure 1."

By providing introductory statements, you again are telling your reader exactly what you are about to describe. If your results section is particularly long, with many different conclusions, these introductory statements make it easier for the reader to see the bigger picture of the section.

It is equally important to add a concluding statement at the end of your paragraphs (as well as at the end of your results subsections in general), thereby providing your reader with a one-sentence recap of the main result. Typically (and ideally), your concluding statements should set the reader up for the next introductory statement they will read in the subsequent section. These introductory and concluding statements can turn your results into a "story" and help the information have better flow.

Examples of results paragraphs with introductory and concluding statements underlined are shown on the next few pages. Note that these paragraphs *could* stand alone without the introductory or concluding statements, but providing these statements to the reader helps with setting the stage for the topic in both this paragraph and the next.

Example 1:

<u>We first evaluated the effect of the combination of IATL and cisplatin on DU145 and PC-3 cell lines.</u> *The MTT assay showed that cisplatin inhibited cell growth as monotherapy, but the inhibitory effect was stronger when in combination with IATL (Figures 1A,B). The combination index (CI) analysis suggested that IATL and cisplatin acted synergistically in DU145 and PC-3 cells (CI<1) (Figures 1C, D). To further investigate the synergistic effect of IATL and cisplatin, the cell apoptosis rate was determined by annexin V/PI staining. The combination of IATL and cisplatin significantly increased apoptotic cell death in both DU145, and PC-3 cells as compared with the agent alone (Figures 1E-H). Furthermore, IATL in combination with cisplatin significantly increased the activity of caspase-3 and caspase-9 (Figures 1I, J).* <u>These data indicate that IATL synergized the chemotherapeutic effects of cisplatin in prostate cancer cells.</u>

Text from: *Huang H, Li P, Ye X, et al (2021). Isoalantolactone Increases the Sensitivity of Prostate Cancer Cells to Cisplatin Treatment by Inducing Oxidative Stress. Frontiers in Cell and Developmental Biology. 9: 632779. Text is from original article. Reprinted under a Creative Commons Attribution License.*

Example 2:

<u>We performed a geospatial analysis combining wildfire extent, burn severity, hydrography, and ecoregion layers to calculate the first estimate of SLBA in different ecoregions in the western US (see Methods).</u>

Wildfires included in the dataset (n = 7677, Supplementary Table 1) span 9 ecoregions in 11 states (Fig. 1). The total SLBA (324,080 km, Supplementary Table 1) represents 5.7% of total stream + river length for the study area, varying between ecoregions from 0.1% (Marine West Coast Forest) to 12.4% (Mediterranean California). Kernel density plots demonstrate considerable latitudinal and longitudinal variability in SLBA, with maxima between 40°N and 45°N, and 110°W and 120°W (Fig. 1). In this area, composed primarily of Cold Desert and Western Cordillera ecoregions, streams are disproportionately impacted by wildfire. Ecoregion-specific density distributions highlight differences in the impact of a single ecoregion on the cumulative spatial density of streams impacted by wildfire. For instance, Marine West Coast Forests, the ecoregion with the lowest wildfire activity (Supplementary Table 1), has clearly defined latitudinal and longitudinal peaks, but does not noticeably alter the general density curve (Fig. 1). In contrast, the similarly sized but more wildfire-prone Upper Gila Mountains ecoregion drives a clear second latitudinal peak near 33°N. <u>Thus, ecoregion-specific parameters drive differences in how wildfire impacts streams.</u>

Text from: *Ball G, Regier P, González-Pinzón R, et al (2021). Wildfires increasingly impact western US fluvial networks. Nature Communications. 12: 2484. Text is from original article. Reprinted under a Creative Commons Attribution 4.0 International License.*

Example 3:

<u>*To further characterize intestinal IETs, dual staining of CD3 protein and T receptor delta constant (TRDC) mRNA was completed in a subset of jejunal, ileal, cecal, and colonic tissues.*</u> *Using the rationale that CD3 protein (yellow staining) would be expressed by all T cells, TRDC mRNA (red staining) would only be expressed by γδ T*

cells, and that TRDC-specific red staining would mask co-localizing CD3-specific yellow staining, cells staining red were presumably γδ T cells (TRDC+), whereas cells staining yellow were presumably αβ T cells (CD3+TRDC−). Staining revealed the presence of both TRDC+ and CD3+TRDC− cells, corresponding to presumable γδ and αβ T cells, respectively, within the epithelium of all intestinal tissues analyzed (Figure 1D). T cells in the lamina propria and submucosa were primarily CD3+TRDC−, with TRDC+ cells noted infrequently outside of the epithelium (Supplementary Figure 2). <u>These findings suggest both αβ and γδ T cells are located within the intestinal epithelium of both small and large intestine in pigs.</u>

Text from: *Wiarda JE, Trachsel JM, Bond ZF, et al (2020). Intraepithelial T cells diverge by intestinal location as pigs age. Frontiers in Immunology. 11: 1139. Text is from original article. Reprinted under a Creative Commons Attribution License.*

Further Reading:

- Bahadoran, Z., Mirmiran, P., Zadeh-Vakili, A., Hosseinpanah, F. & Ghasemi, A. The Principles of Biomedical Scientific Writing: Results. Int. J. Endocrinol. Metab. 17, e92113 (2019)

Example header from:

- Boettcher AN, Li Y, Ahrens AP, et al (2020). Novel Engraftment and T Cell Differentiation of Human Hematopoietic Cells in ART (-/-) IL2RG (-/Y) SCID Pigs. Frontiers in Immunology. 11: 100. Reprinted under a Creative Commons Attribution License.

8
INTRODUCTION

As indicated by its name, the introduction of your manuscript or report introduces your readers to the topic you are going to present. If a reader decides to read the introduction of your paper first (as opposed to reading the discussion section first, which some readers may do), you want your introduction to be enticing enough that the reader will become engaged in the content and want to read more.

The most important aspect of the introduction is to provide enough context so that the reader can understand why you performed the presented research. In this chapter, I propose four main questions that you should answer when writing your introduction. To walk through each of these questions, I have included example texts from three different published manuscripts (A, B, and C; citations can be found at the end of this chapter). In all of these example texts, answers to all four questions have been provided within the introduction. From these examples, you can see how the information can be smoothly incorporated into an introduction.

1. What does your reader need to know about the research field?

Typically, the first paragraph of an introduction provides a broad overview of the specific field that the research falls into. In composing this first paragraph, it is important to determine what overall background information a reader might need to know to understand the context of your research. First paragraphs typically only contain general information.

Example A:

Multiple myeloma (MM) is a plasma cell malignancy that arises from a single clone of malignant plasma cells (PCs) in the bone marrow. It has a worldwide incidence of 2.1/100,000 new cases every year, ranking as the second most common hematological cancer[1]. MM has usually a late onset with a mean age at diagnosis around 60 years[1].

Example B:

Chronic rhinosinusitis (CRS) is a disabling disease affecting more than 10% of the general population (Hastan et al. 2011). CRS is classified into CRS with nasal polyps (CRSwNP) and CRS without nasal polyps (CRSsNP) (Fokkens et al. 2020).

Example C:

Ovarian cancer (OvCa) is the most lethal among gynecologic malignancies, taking an estimated 14,000 lives in the United States in 2018 (1). OvCa often goes undetected until late stages due to non-specificity of its early symptoms, hence 2/3 of patients have late-state disease (stage III–IV) at diagnosis. The current standard of care is debulking surgery to remove tumor masses followed by first-line platinum and Taxol chemotherapy (2). Debulking is critical to successful chemotherapy, and so prior identification of tumor masses by diagnostic imaging often plays a key role in pre-surgical planning. X-ray computer tomography (CT) (3) is the most widely used imaging modality for evaluating peritoneal spread in OvCa for presurgical planning, yet there are well acknowledged "blind-spots" where tumor spread simply cannot be seen as the contrast between normal tissue and tumors is insufficient to discriminate one tissue type from another. Thus, tumors can be missed, leading to incomplete tumor resection.

2. What background information does the reader need to know to understand the research?

After you present the general information about the field of study, you can use the next paragraph to provide specific information for your reader. This specific information sets the stage for when you present the problem your research addressed. For example, if you worked on developing a new PET tracer for imaging a rare form of sarcoma, it would be worthwhile to describe the mechanism for how the tracer will target the cells (antibody-mediated binding to a surface receptor, cellular uptake of a metabolic analog, etc.). In cellular and biological studies, describing mechanisms of action or relevant pathways may be important to include in your introduction.

As another example, if your study is describing the synthesis of a new molecule, you should describe the components used in any relevant prior studies related to the synthesis procedure.

Here are some other examples:

Example A:

Alongside polymorphic variants, another emerging marker of susceptibility and prognosis for several diseases is telomere length. Telomeres are specialized structures that cap the end of chromosomes and are involved in several processes like cellular growth, chromosomal stability, and proper segregation to daughter cells[16]. Telomerases are the most important enzymes involved in telomere replication and their dysfunction is considered a cancer hallmark[17]. Telomere length measured in leukocytes (LTL) has been used to investigate the susceptibility to different cancer types including several hematological malignancies [18,19,20,21,22,23,24,25], among which MM for which only two studies of small size have been attempted, a prospective cohort study and a retrospective case control study [21,26].

Example B:

Epithelial cells, connective tissue cells and immune cells contribute in various ways to the pathogenesis of CRS.

Example C:

Current research to improve imaging technologies and methodologies uses either human volunteers or rodent models. Imaging research involving human OvCa patients is challenging for a variety of reasons. For imaging modalities which involve significant radiation, such as CT, extensive research cannot be performed due to radiation dose. Second, it is challenging to perform serial imaging research on OvCa patients due to ongoing treatment regimens and patient morbidity. Such serial scans could be useful in developing predictive imaging capability derived from a multiparametric data set (11).

3. What is the main problem that you are trying to address?

It is important to set the stage for your research by telling the reader specifically what your research is going to address. If you do not explicitly state the problem, a reader may be left wondering why this particular study was done. If a reader does not understand the motive for a study, they may be deterred from reading more of your paper. Therefore, you need to make sure to explicitly state the underlying motive and problem that you have addressed with your study.

For example, maybe your study was about optimizing the synthesis of a chemical because other commonly used methods require long processes or are too expensive. In this case, you may need to explain to your readers why the current methods are not optimal and how the research you are about to present is going to improve the synthesis methods.

As a different example, let us say that you sequenced a variety of patient tumors to better understand disease progression and response to therapy, as well as to provide researchers with a new genetic resource. A key aspect that you could mention in your introduction is that the availability of a new genetic resource can foster more research, while also helping other researchers gain insight on certain patient populations.

Below are excerpts explaining the problem that will be addressed within each study.

Example A:

Recently genome-wide association studies (GWAS) have identified 11 SNPs [23,34,35,36] that, collectively, explain 2.28% of LTL variability and can be used as a genetic instrument in a Mendelian randomization fashion. This approach has been already successfully tested in relation to the risk of developing various cancer types [23,37,38,39,40,41,42,43,44,45], but was never done in MM.

Example B:

However, few data on the relative abundance of these cell lineages are available (Jiang et al. 2013).

Example C:

Development of imaging strategies on rodents poorly informs how one approaches the clinical scenario. This is due to the size of the animal and the tumors, as well as the equipment that are used in animal imaging experiments. Small tumors exhibit different perfusion (12) and water diffusion (13) from large tumors. Small animals have different metabolism than large animals and exhibit vastly different pharmacology of administered drugs (14), while humans and pigs have similar liver content of cytochrome proteins (15, 16). (...) Recently there is a new hybrid field of molecular imaging and surgery called optical surgical navigation (17). This field couples fluorescence imaging with surgery to enhance surgical removal of tumors by way of fluorescent marker uptake. The translational value of new fluorescent tracers, either targeted or untargeted, can only be meaningfully evaluated in the context of a research subject that has appropriate size and physiology to OvCa patients. A pig model of OvCa could fill this crucial gap.

4. What were the hypotheses or main questions you had when you started your project?

Lastly, in the process of crafting your research story, you need to tell your reader what your original study hypotheses or questions were. There may be some cases when a research project was exploratory in nature, with no specific hypotheses, in which case, a hypothesis does not need to be presented. By sharing your original hypothesis or hypotheses, you set the foundation so that your reader can start thinking about the types of specific research questions and experiments you will talk about in your manuscript.

See a few examples below.

Example A:

We have tested the genetic score (henceforth called "teloscore") in the context of the International Multiple Myeloma rESEarch (IMMeNSE) consortium to investigate the relationship between gdTL and MM risk and survival.

Example B:

Our aim was to provide quantitative, purely informative data of the distribution of these cell lineages and their coexpression patterns in healthy and diseased upper airway mucosa. Coexpression patterns might identify, e.g., cells in the epithelium undergoing through epithelial-mesenchymal transition (EMT) (Li et al. 2019; Yan et al. 2019). We investigated whether immunofluorescence multichannel image cytometry (IMIC) provides such data.

Example C:

To initiate the development of a large animal model of ovarian cancer, we tested whether human ovarian cancer cells could survive and develop ectopic tumors in SCID pigs. OSPC-ARK1 cells, derived from an ovarian serous papillary carcinoma (OSPC), were injected into male and female SCID pigs and were monitored for tumor development for this first stage screen.

Bringing all the questions together: the hourglass

Deciding how to present information—and in what order—within the introduction can be tricky. I like to think of the entire manuscript as an hourglass, the introduction being the top part of the hourglass. In your introduction, you should start broad; this can help pique a reader's interest, even if they may not know a lot about the field. As you progress through the introduction, the information that you present becomes more and more detailed as you describe the specific problem areas that your study addresses. The end of the introduction contains the most specific content, as this is where you will provide readers with your study's hypothesis and questions, right before heading into the methods or results sections.

Depending on the journal, you may have space at the end of the introduction to provide two to five sentences about the main results of your study. Not all scientific fields may write this way, but it is an option. By providing an overview of the main results in the introduction, the reader will already have a clear idea for how to interpret the results that you present in your study.

The following page shows the top of the hourglass and the organization for your introduction. We will revisit this hourglass in Chapter 10 for the discussion section.

1. Provide overview of the system being described

2. Describe the broader implication of the question you are addressing

3. Describe main issue being addressed

4. Present study hypotheses

5. Describe general methods and findings of your study

.
.
.

Methods & Results

Further Reading:

- Bahadoran, Z., Jeddi, S., Mirmiran, P. & Ghasemi, A. The Principles of Biomedical Scientific Writing: Introduction. Int. J. Endocrinol. Metab. 16, e84795-e84795 (2018)

Cited examples:

Example A:
- Giaccherini M, Macauda A, Orciuolo E et al (2021). Genetically determined telomere length and multiple myeloma risk and outcome. Blood Cancer Journal. 11: 74. Text from original article; original in-text citations were retained. Reprinted under a Creative Commons Attribution 4.0 International License.

Example B:
- Giotakis AI, Dudas J, Glueckert R, et al (2021). Characterization of epithelial cells, connective tissue cells and immune cells in human upper airway mucosa by immunofluorescence multichannel image cytometry: a pilot study. Histochemistry and Cell Biology. 155: 405-421. Text from original article; original in-text citations were retained. Reprinted under a Creative Commons Attribution 4.0 International License.

Example C:
- Boettcher AN, Kiupel M, Adur MA, et al (2019). Human ovarian cancer tumor formation in severe combined immunodeficient (SCID) pigs. Frontiers in Oncology. 9:9. Text from original article; original in-text citations were retained. Reprinted under a Creative Commons Attribution License.

9
MATERIALS AND METHODS

The methods section is a vital part of your manuscript. As readers are working through your report or manuscript, they will need to know how certain aspects of the study were done. It is your job to give the readers an accurate account of how you performed the experiments and trials.

During peer review, major revisions may be required for manuscripts that have not described the methodology clearly enough. If your methods are not clear, the editors and peer reviewers may not be able to understand what you did, which can in turn make it difficult for them to interpret the results of your study.

As you review your methods, try to read them as an outsider—as someone who was not in the lab—to see if the steps you wrote make sense and are presented in a coherent order.

Write down your methodologies as you perform experiments

There may be months or even years that elapse between the day you perform an experiment and the day you decide to start writing. For that reason, you will need to take detailed notes on all aspects of your experiments: protocols, solutions used, solution concentrations, incubation periods, product names and manufacturers, animal numbers, doses, and other relevant information. If you do not take detailed notes, you may forget the specifics of what you did, and it could take you a long time to recover the information again. To prevent this from happening, a best practice is to take detailed notes as you are performing the experiment and then save them in a place where you can easily find them again.

In clinical studies, make sure you document criteria that were used for participant eligibility and inclusion. Similarly, ensure that you track the number of participants that were excluded as well as the reason for the exclusion. For example, if you developed a deep learning model to segment and stage brain tumors on MRI in children, it would be important to indicate the age ranges of the patients that were used for model development. In this example, such a model may not perform well on adults.

If you have repeated an experiment multiple times and have found that your results are repeatable (i.e., you obtain the same results each time), it is especially important to maintain a record of the specific methodologies that were used. You may even consider writing a draft of the methodologies section at the time so that you will have it readily available when you begin to put together a report or other manuscript in the future. Remember that the writing does not need to be perfect in the first draft, but by simply creating the document, you already have something that you can use when you start to write your manuscript in earnest down the road.

Methods should mirror the results section

Because a reader may read through your entire paper (starting from the introduction), rather than just jumping straight to the results, it is best to write your methods in the same order in which you will present the associated results. For human and animal studies, patient populations and animal models should be described first. For chemical engineering studies, you can start off by describing the synthesis of the molecule you will describe in the paper. Typically, the first few subsections of the methods will describe the model or main subject of investigation. Subsequent subsections can then go into detail about the specific experiments that were performed. Typically, the last part of the methods should describe the statistical analysis.

Split your methods into subsections based on experiment

Use headers in your methods if the journal you are submitting to allows. I have seen many papers where the materials and methods section was presented as a large block of text with no headers or indications as to what each paragraph is about. Without detailed headers, it can be difficult for readers to quickly identify methodologies that they may be interested in. Your headers can be short (three to eight words), but they should be descriptive. The following are some examples:

- RNA extraction from tissues
- Flow cytometric analysis of lymphoid organs
- MRI acquisition
- Mouse models
- Tumor inoculation with MDA-MB-231
- Statistical analysis
- Convolutional neural network training
- Study participants
- Nanoparticle synthesis

Add in regulatory approvals

All studies that involve human or animal subjects require ethical approval prior to the study being initiated. Typical oversight committees include the Institutional Review Board (IRB) for human studies and the Institutional Animal Care and Use Committee (IACUC) for animal studies. Depending on your institution, the naming conventions of these committees may vary. Journals require statements regarding ethical approvals in studies that involve human or animal subjects, so be sure to include these statements within your manuscript. For human subject research, you should also include a statement on Health Insurance Portability and Protection Act (HIPAA) compliance (in the United States) and note if participants had to give written consent or if written consent was waived.

Clearly define your statistical tests

Running and reporting on proper statistical tests for your datasets is essential. While this chapter does not detail specific statistical tests, I have provided a list of references at the end of this chapter for further reading on different statistical tests that may be relevant for your data. In your methods section, you should clearly state the statistical tests that were used and which specific hypothesis they were used to test. Many issues can arise during the peer review process if an improper statistical test was performed. For example, use of an incorrect statistical test may have led you to conclusion that may not be properly supported. Consult with a statistician to assess which tests to use for your datasets. If you are working in a research environment, it is likely that your university, institution, or company has a statistical expert on staff.

Further reading:

- Levine, D., Bankier, A. A. & Halpern, E. F. Submissions to Radiology: Our Top 10 List of Statistical Errors. Radiology 253, 288-290 (2009).
- Mishra, P. et al. Descriptive statistics and normality tests for statistical data. Ann. Card. Anaesth. 22, 67-72 (2019).
- Evans, S. R. Common statistical concerns in clinical trials. J. Exp. Stroke Transl. Med. 3, 1-7 (2010).
- Ghasemi, A., Bahadoran, Z., Zadeh-Vakili, A., Montazeri, S. A. & Hosseinpanah, F. The Principles of Biomedical Scientific Writing: Materials and Methods. Int. J. Endocrinol. Metab. 17, e88155 (2019).

10
DISCUSSION

The main point of the discussion section is to emphasize the impact of your research. The discussion is a place where you can recap your data, put your results within the context of the field, and discuss future outlooks. In addition, in the last paragraph of the discussion section, it is important to provide a succinct overview of your study; this overview will consist of the main original question that was proposed, the key results you found, and the broader impact of the finding.

This overview is important for two reasons: (1) to remind your readers again what the point of the study was and (2) to provide this information to readers who may read the discussion portion of the paper first.

Some journals may have a format in which the results and discussion are combined. For these journals, listing out the results once—at the first mention—is sufficient. You can mention limitations as they come up in your data presentation. For example, you could tell your reader about a specific limitation after presenting the results for a particular experiment. In manuscripts that have a combined results and discussion section, it is important to make sure that the concluding paragraph of the entire document still contains a broad overview of the findings of the study.

Outlined throughout the rest of this chapter are questions to address within your discussion. Three examples (Examples A, B, and C) are shown for each question with text that the author provided to address these questions. The citations for all three examples are at the end of this chapter.

1. What are the main key findings?

In the first paragraph of your discussion, you should remind your readers of what the original main research questions were. Recap these questions and objective within the first one or two sentences of the first paragraph. In the next four to six sentences, recap the main key points of your study. Think back to the key points that you developed for your results section—you can use that information again here. If you had a longer results section and a reader was focused in on the details, a summary at the beginning of the discussion is a nice way to bring them back into the larger picture of your study. Below are some excerpts of main key findings from different studies.

Example A:

Our results recover three main clades (2, 9 and 13). The monophyly of these lineages and their relationships are strongly to moderately supported, representing substantial progress in the establishment of a robust evolutionary tree for the genus Exocelina. Several morphological species-groups (MSG) appear paraphyletic or polyphyletic suggesting the need for a revision of species-group composition and key synapomorphic characters.

Example B:

The acylation of highly aggregated and low soluble industrial side-stream PPC using EDTAD resulted in an increase of the swelling capacity of the PPC from 2 to 24 g g−1 (2400%).

Example C:

In this work, we focused on the evolutionary potential and the genetic basis of bacterial adaptation to biocides and their cross-resistance to antibiotics. Our results suggest that biocides invoke disparate evolutionary trajectories, with some promoting the emergence of cross-resistance, while others do not. Chlorophene, benzalkonium chloride, chlorhexidine, and glutaraldehyde were the top biocides in the first category.

2. How do your results fit into previous research in the field?

Next you should engage in a discussion about how your results contribute to your specific scientific field. Did your research reveal a new mechanism that was previously unknown? Did you develop a model that had higher performance than others that had been developed in the past? Did you optimize a protocol that produces more accurate or faster results?

It is essential to provide literature support within your discussion. I have edited manuscripts whose discussion sections contained a 1000-word (or more) review of their results (with no citations), which is not the purpose of a discussion section. Think of what you might want to say to someone if they asked you how and why your research is important to the field. To answer this effectively, you would need to provide some type of comparison of prior works. To that end, be sure to include such comparisons in your discussion.

Example A:

These observed discrepancies are largely due to the methodology used to obtain absolute divergence times using molecular clock calibrations. Earlier studies mostly relied on beetle rates of evolution or fossil calibrations placed in the outgroups, while the current one relies on secondary calibrations applied across a large selection of lineages and derived from a robust fossil-based evolutionary framework of diving beetles [66]. Hence, we believe these new estimates to be much more in line with the evolutionary history of these diving beetles than earlier estimates. Based on these new estimates, we infer that the three main clades 2, 9 and 13 diverged in the mid-Miocene ca. 15 Ma. The rest of diversification was gradual up to the last million years within the Pleistocene.

Example B:

Although the protein aggregation in the PPC decreased the acylation efficiency, the endogenous cross-links of the industrially processed protein yielded a stable hydrogel without the need for toxic cross-linkers, which represents an advantage compared to other proposed protein-based SAPs.

Example C:

In contrast to existing literature (Kampf, 2018), we report on the cross-resistance potential for exposure to glutaraldehyde.

3. Based on the new information gained from your research, what are some follow-up questions?

As you start to think about the impact of your work on the field, you may find that there are important follow-up questions that arise. You likely have already thought about many of these questions if you are already working on a next set of experiments. For example, if you worked on the development and characterization of a new immunotherapeutic that has shown efficacious for reducing tumor growth, a next question may be to address the mechanism of action. If you described the safety profile of a new CT contrast agent, a next step may be to move towards testing it in humans.

In writing about these next steps, you do not have to spill all your secrets. The purpose of providing next steps in the research path is to show your readers that your work is not a dead end. Additionally, providing follow-up questions makes it more likely that you will keep your readers engaged and interested in your work.

Example A:

On lower elevations, species of the closely related genus Copelatus commonly occupy the same microhabitat as Exocelina. With Copelatus species being larger than Exocelina, competition between the two in sympatry may limit the potential number of Exocelina species in such habitats [46]. However, this remains to be tested with additional data, and thus, our results need to be carefully interpreted in light of the current debate regarding the potential pitfalls when estimating diversification rate dynamics from extant phylogenies [67].

Example B:

This result opens-up opportunity for future optimisation work in the acylation of the PFJ before the severe industrial treatment to produce PPC, which could result in greenhouse emission savings compared to the acylation process using PPC. The high blood and water swelling combined with the moisture uptake of the acylated PPC make the material a promising candidate for applications where today synthetic SAP are the material of choice, e.g., in sanitary pads and desiccants.

Example C:

Two of the evolved strains had a considerable increase in biofilm formation: pov 1b/2a and glu 2b. For pov 1b/2a, such effect can be attributed to a mutation in envZ given the known role of the gene for the regulation of flagellar and curli genes (Brombacher et al., 2003; Prüß, 2017). However, the biofilm increase in glu 2b could not be easily explained based on the current literature and suggests a connection between the genes affected (yeaW and pyrE) and the phenotype observed. The repair of each gene mutation individually partially restored the non-biofilm forming phenotype, suggesting a synergic relationship between the gene products. Further experiments would be needed to clarify such a link.

4. What were the limitations of your research?

It is important that you acknowledge the limitations of your research. At some point during the project, a decision was made to stop conducting experiments and move, instead, towards publishing. It is likely that there is something within your experimental design that could have been optimized or added to help draw even more concrete conclusions. For example, let us say that we developed a deep learning model for segmenting glioblastoma lesions on brain MRI. In the study, we used only data from one institution to develop the model. Thus, a key limitation in this study would be that the data came from a single site as opposed to multiple sites. Obtaining data from multiple sites is not trivial task.

Limitations are not barriers to publication, but they are important to mention so that readers have a full scope of the balance between the findings you have presented and the limitations of the study. In identifying these limitations, you may also be able to add some follow-up questions or objectives for future studies, as mentioned in the previous section. Additionally, reviewers will often ask for limitations if they are not provided. Thus, providing them ahead of time—in the submittal draft of a manuscript—may satisfy reviewers and result in less work for you during the revision process.

Below are examples of limitations.

Example A:

While our ultimate aim of employing this technique was to determine if there are spatial patterns within our database, we did not apply a spatial constraint to the algorithm, which means that the algorithm assumes there is no spatial correlation between any two sample locations. This means that if a spatial pattern is identified within the analysis, it is dependent on the altitude or underlying geology (or both). The tool requires the user to specify how many populations between 2 and 15 might exist within the dataset. After running the algorithm using different input parameters, we limited the final output to five populations. Readers should also note that it is possible that the numerical value assigned to represent the altitudinal range and geological terrane might also influence the results determined by this algorithm. Considering these limitations, we used the tool for data exploration purposes only, using it to test if any apparent relationship could be identified between species and altitude, species and underlying geology and species plus both altitude and geology.

Example B:

See excerpt from question 2 from this paper.

Example C:

We limited our study to a single bacterial species in a laboratory environment to better compare biocides and control test conditions. In the field, however, other mechanisms for resistance emergency and maintenance in bacterial populations may be at play.

5. Can you succinctly summarize your findings with potential next steps?

The very last paragraph of your discussion should provide a summary of the entire study. Think about what information you would want to provide if you knew a reader was going to read the very last paragraph of your paper first. If your manuscript combined the results and discussion together, providing just a statement on the next potential steps is sufficient for your concluding statement (as shown in Example B below).

Example A:

We identify colonization events from the Central Range East towards the Papuan Peninsula from the mid Miocene, and then repeated interactions with the Central Range and north coast areas. (Fig. 6). The Bird's Head might have been colonized earlier than expected based solely on geological evidence, but major lineage diversification appears to be more recent, from ca. 5 Ma (this study; [42]; Table 4). Here, studies more focused on the Bird's Head, with comprehensive taxon sampling, remain a future task.

Example B:

Future work should focus on the industrial scalability of the method, especially in the area of unreacted EDTAD recovery and material performance over time in storage.

Example C:

Biocides are of fundamental importance to control and eliminate pathogens in high-risk settings such as hospitals. Our work highlights the need to stimulate the responsible use and discard of such products to avoid unnecessary bacterial exposure and potential selection pressure for cross-resistant strains. In addition, we acknowledge that further analysis of other bacterial species and clinical isolates of E. coli, as well as mixed-species communities, to multiple biocides in a similar fashion of this work have the potential to amplify the knowledge provided here.

Brining all the questions together: the hourglass

Let us jump back to the hourglass structure that we discussed in Chapter 8. In the beginning of your discussion section, the content will be highly specific, as you will be presenting a review of the data and results. However, as you progress through the paragraphs of your discussion, the scope of the content should broaden as you conclude your paper. The last one to two sentences of your paper should frame the conclusions of your work within the broader context of the scientific field. Make sure that you do not show specific data within the last paragraph of the discussion. Instead, provide your readers with statements about the study's conclusions. The hourglass for the discussion section is shown on the next page.

Methods & Results
.
.
.

1
Recap of original questions and study findings

2
Discuss how your results contribute to the field and compare to other studies

3
Discuss the limitations of your study

4
Describe some next questions that could be addressed

5
Provide succinct overview, with broader context and impact to the field

Further Reading:

- Jawaid, S. A. & Jawaid, M. How to write introduction and discussion. Saudi J. Anaesth. 13, S18-S19 (2019).
- Masic, I. How to Write an Efficient Discussion? Med. Arch. (Sarajevo, Bosnia Herzegovina) 72, 306-307 (2018).
- Höfler, M., Venz, J., Trautmann, S. & Miller, R. Writing a discussion section: how to integrate substantive and statistical expertise. BMC Med. Res. Methodol. 18, 34 (2018).

Cited Examples

Example A:

- Toussaint EFA, White LT, Shaverdo H, et al (2021). New Guinean orogenic dynamics and biota evolution revealed using a custom geospatial analysis pipeline. BMC Ecology and Evolution. 21: 51. Text from original article; original in-text citations were retained. Reprinted under a Creative Commons Attribution 4.0 International License

Example B:

- Capezza AJ, Muneer F, Prade T, et al (2021). Acylation of agricultureal protein biomass yields biodegradable superabsorbent plastics. Communications Chemistry. 4: 52. Text from original article; original in-text citations were retained. Reprinted under a Creative Commons Attribution 4.0 International License

Example C:

- Pereira BMP, Wang X, and Tagkopoulos I (2021). Biocide-induced emergence of antibiotic resistance in escherichia coli. Frontiers in Microbiology. 12:640923

Text from original article; original in-text citations were retained. Reprinted under a Creative Commons Attribution License

11
FIGURES AND TABLES

Figure and table creation is an art form within itself. When you start putting these data compilations together, consider designs and layouts that will make it easy for readers to understand the main points of the data. As we perform experiments for a study, we produce a lot of data. It may not be necessary to show every single piece of data collected. This chapter will provide some general guidelines for making figures and tables.

A figure and table checklist

Figure and table design will vary depending on your field and the experiments you performed.

Listed below are general topics to consider as you are designing your figures and tables. This is not an exhaustive list; however, being aware of these particular topics will help you make figures and tables that are easier for readers to digest.

- Table legends
  - Legends provide additional information that may not necessarily fit within a table. For example, you may need to define how data are presented within a table. A value of 56 (43-69) could be a mean (range) or median (interquartile range). Make sure you indicate these definitions in your legend. You may also need to add information relevant to the statistical tests used and P values used for significance. A legend may also have a list of expanded abbreviations that were used within the table.

- Figure Legend
  - The first sentence of the figure legend should typically be a one-sentence summary of the main point of the entire figure.
  - Next, if your figure contains multiple panels, make sure you provide a description for each individual panel. The text content within the figure legend should describe the method or experiment that was performed to obtain the data, and not the results from the experiment.

While you should not list your results in the figure legend, you can show P values for specific comparisons from data presented within the table or figure. Descriptions of the results, though, should only be in the main text. Thus, you need to ensure that there is enough methodology detail in the figure legend and clarity within the figure itself that a reader can understand the associated data.
- Lastly, make sure to indicate any statistical analyses that were performed on the data being presented and the P values that were used for defining statistical significance.

- Abbreviations
  - It is best to have a low number of abbreviations in a manuscript because readers may have a difficult time keeping track of them. However, when considering wording within tables and figures, the space you have available is also an important factor. In tables and figures, it is acceptable to show abbreviations that readers may be less familiar with simply for the sake of saving space. Tables and figures that are crammed with unnecessary words may look unorganized and are difficult to read. Chapter 15 has some general rules regarding abbreviations within the main text of a report or manuscript.

- Units
  - Always define your units in tables and on figure axes. If you do not write down the units, the reader will not know what you are showing.

- Statistics
  - If the data you are presenting within a table or a figure has associated statistical tests, make sure to include the statistical test and *P* values within the figure legend. If there are numerous comparisons being made, you can use asterisks to indicate different *P* value thresholds. Also make sure to add the *n* for the sample size.

- Colors and labels
  - If you have graphs or charts with multiple variables, try to use colors or symbols so that readers can easily tell which group is which. Be sure to include a key within the figure itself. I have sometimes seen descriptions only mentioned within the figure legend (e.g., blue bars indicate group A and gray bars indicate group B), but it best to include this within a key in the figure itself as well. Showing this information within the figure makes it much easier for readers to identify what they are looking at. Additionally, these keys and colors

are helpful to readers who may look at the figures before reading any of the text.

- Number of figures and tables
  - Follow journal figure and table limits. Showing every single piece of data you have in the main text of a paper is not always necessary, nor is it easy for a reader to digest. Only show the data that is needed to convey your main key results. The next section discusses what to do with your extra data.

- Supplemental data
  - Most journals will allow you to submit supplemental data with your study. In some of your experiments, you may have obtained data that was important, but may not contribute as much to the results of your paper as some of your other experiments do. For example, if you were testing a drug in a tumor cell line and the drug led to a reduced growth rate and downregulation of proteins involved in the cell cycle, this may be one of the main results that you would include in the main text. If you found that the drug also upregulated immune-related proteins, but this is not as important for your main conclusions, you could add this data within the supplement material.

- Avoid non-data tables and figures within the main text
  - In original research papers, it is best to forego the inclusion of any figures and tables that do not directly show new data from your study. I have often seen figures and tables used in introductions to provide examples or overviews of previous findings. However, these can be distracting to the reader if they are looking for the results of your paper and are instead met with non-results content. These types of figures and tables could be mentioned within the materials and methods section and be placed within the supplemental material instead. Try to include only new data relevant to your study within your tables and figures.
  - A couple of exceptions to this rule include figures showing an overview of methodologies or summary figures showing a new mechanism or process.

12
TITLE AND ABSTRACT

The title and abstract are both key components for grabbing the attention of potential readers. As such, the title needs to reflect the main area of study and conclusion of your paper, and the abstract needs to be a detailed, yet succinct, summary of the following:

- the problem being addressed,
- what you did and how you did it,
- the results, and
- a general conclusion.

Key word indexing

Both the title and the content of your abstract

will be used for the indexing of your article. Imagine that you are a prospective reader of your article—what key words would someone need to enter into a search engine for your study to show up? Add these words into your title and abstract so that your article has increased visibility in search engines.

You may need to do some research to determine which key words are optimal for your particular study. For example, if your study involves the optimization of an immunohistochemical panel for staining tumors, you may want to emphasize the specific species that it was performed on (i.e. adding "murine", "rat", "pig" or other species within the title and abstract), as this could be an important aspect to prospective readers.

Additionally, there may be some terms that could be presented in different ways (e.g., PCa vs prostate cancer; PCA vs principal component analysis; or ATP vs adenosine triphosphate). If you have phrasing within your title that could be presented in more than one way, it is a good idea to input the different terms into a relevant academic search engine to see which phrasing results in the higher number of results. For example, on the day this particular paragraph was written, a search of "deoxyribonucleic acid" on PubMed led to 6,397 results, whereas a search of "DNA" led to 721,259 results. With these particular results, you should consider using the term "DNA" in your title and abstract, as the search engine results indicate that others in the field likely use this phrasing more frequently when searching for papers.

Abstract structure

Your abstract will be the first text that someone will read to decide if they want to see the rest of the paper. Therefore, it is important that you describe all important aspects of your paper concisely.

Back in Chapter 7 we discussed identifying the main key points to plan out your results section. Those same main key points should be represented within your abstract too. Abstract lengths can range anywhere from 200 to 500 words, depending on the journal. Composing the abstract after you have finished the rest of the paper may make this section a little easier to write.

As you write your abstract, aim to have one to four sentences for the background, one to three sentences on the methods used, three to five sentences about the results, and then one to two sentences for your conclusion. Breaking the abstract into these separate sections and then combining the sentences together will make the writing process feel more manageable. Some journals will require that the abstract be broken up with headers, while other journals may allow full paragraphs. Read other abstracts from the journal(s) you are interested in to see how others have written their abstracts.

Below are two examples of abstracts with clearly defined sections; the citations for these two articles are found at the end of this chapter.

Example 1 (161 words)

Background and problem:

Vascular-targeted PDT (vPDT) has produced promising results in the treatment of many cancers, including drug-resistant ones, but little is known about its efficacy in lymphoma. Unfortunately, the lack of a specific therapeutic target and a hypoxic microenvironment for lymphoma jeopardizes the efficacy of vPDT severely.

Methods:

In this study, we designed a lymphoma tissue factor-targeted "O2-evolving" strategy combining PDT with catalase and HMME-encapsulated, EGFP-EGF1-modified PEG-PLGA nanoparticles (CENPs) to boost PDT efficiency; this combination takes advantage of the low oxygen tension of lymphoma.

Results:

In our results, CENPs accumulated effectively in the vascular lymphoma in vivo and in vitro, and this accumulation increased further with PDT treatment. Per positron emission tomography imaging, combining CENPs with PDT inhibited lymphoma glucose metabolism significantly. The expression of hypoxia-inducible factor (HIF)-1α in the entrapped catalase groups reduced markedly.

Conclusion:

These data show that the combined administration of PDT and CENPs can prompt tissue factor-cascade-targeted and self-supply of oxygen and that it has a good therapeutic effect on malignant lymphoma.

Example 2 (285 words)

Background and problem:

Breast cancer represents the number one cause of cancer-associated mortality globally. The most aggressive molecular subtype is triple negative breast cancer (TNBC), of which limited therapeutic options are available. It is well known that breast cancer prognosis and tumor sensitivity toward immunotherapy are dictated by the tumor microenvironment.

Method and results:

Breast cancer gene expression profiles were extracted from the METABRIC dataset and two TNBC clusters displaying unique immune features were identified. Activated immune cells formed a large proportion of cells in the high infiltration cluster, which correlated to a good prognosis. Differentially expressed genes (DEGs) extracted between two heterogeneous subtypes were used to further explore the underlying immune mechanism and to identify prognostic biomarkers. Functional enrichment analysis revealed that the DEGs were predominately related to some processes involved in activation and regulation of innate immune signaling. Using network analysis, we identified two modules in which genes were selected for further prognostic investigation. Validation by independent datasets revealed that CXCL9 and CXCL13 were good prognostic biomarkers for TNBC. We also performed comparisons between the above two genes and immune markers (CYT, APM, TILs, and TIS), as well as cell checkpoint marker expressions, and found a statistically significant correlation between them in both METABRIC and TCGA datasets. The potential of CXCL9 and CXCL13 to predict chemotherapy sensitivity was also evaluated. We found that the CXCL9 and CXCL13 were good predictors for chemotherapy and their expressions were higher in chemotherapy-responsive patients in contrast to those who were not responsive.

Conclusion:

In brief, immune infiltrate characterization on TNBC revealed heterogeneous subtypes with unique immune features allowed for the identification of informative and reliable characteristics representative of the local immune tumor microenvironment and were potential candidates to guide the management of TNBC patients.

Further Reading:

- Tullu, M. S. Writing the title and abstract for a research paper: Being concise, precise, and meticulous is the key. Saudi J. Anaesth. 13, S12-S17 (2019).
- Ruffell, D. Writing a great abstract: tips from an Editor. FEBS Lett. 593, 141-143 (2019).

Example 1:

- Abstract from: Li Z, Yin Y, Jin W, et al (2020). Tissue Factor-Targeted "O2-Evolving" Nanoparticles for Photodynamic Therapy in Malignant Lymphoma. Frontiers in Oncology. 10:524712. Text is from original article; section descriptions were added to show the breakdown of the abstract. Reprinted under a Creative Commons Attribution License.

Example 2:

- Abstract from: Lv Y, Lv D, Lv X, et al (2021). Immune Cell Infiltration-Based Characterization of Triple-Negative Breast Cancer Predicts Prognosis and Chemotherapy Response Markers. Frontiers in Genetics. 12:616469. Text is from original article; section descriptions were added to show the breakdown of the abstract. Reprinted under a Creative Commons Attribution License.

13
SELF-EDITING: GENERAL CONCEPTS

The editing process is a critical step when composing a written work. When I edit, I prefer to print my paper out and edit the physical copy. My goal when editing is to make sure that the sections flow and that I have provided enough detail so that the content can be understood by others. I look for any areas that may be confusing and would benefit from rewording.

You may like to edit on physical printed copies (this is what I usually do), but some of you may prefer editing on a computer instead. Other editing methods may include listening to your written piece with a text-to-speech program or editing on a touch screen on a tablet allowing you to "write" on the document. You will need to find which method works best for you. No matter which method you choose, a general best practice would be to pace yourself. After you finish editing a section, leave it but then come back to it a day or two later. This break allows you to look at the section with a fresh set of eyes, and you may identify additional places that could benefit from more editing.

Additionally, another aspect to consider as you edit your work is that a prospective reader may start reading your paper at any section. In other words, some readers may read the discussion first, while other readers may jump straight to the figures. Therefore, it is important to have your data and points clearly presented in each section to keep your reader engaged. The more engaged the reader is, the more they will read!

Be aware of self-plagiarism

All text within a manuscript should be original; when you write, make sure you put your thoughts in your own words. It is important to ensure that you have not plagiarized any material from previous works, even from your own prior work. During the submission process, journal staff will assess your document for any plagiarized material. If there are substantial areas of plagiarism within your manuscript, it will be sent back to you as either a rejection or for revision, and you may end up jeopardizing your opportunity to publish in a specific journal. Depending on your university or workplace, you may have access to a free plagiarism checker software. Consider using one of these programs prior to submitting in order to double-check that your manuscript contains all original text.

Self-editing checklist

1. Confirm consistency in data and methodology across the entire manuscript

- Do you have your data or methodology written out in your abstract?
- Is the information consistent with the information within the main text?
- Does the data you have written in the results section match what is in your figures and tables?

Sometimes inconsistencies can pop up when you are writing, e.g., changing a "74" to a "47" or writing out the wrong numerical information for a given experiment. Before you submit your paper, make sure your methods and results are consistent across all parts of the manuscript. If there are inconsistencies, it can make it difficult for reviewers or editors to assess your work, as they will not know which numbers are correct. Additionally, for any type of statistical analysis that you perform, make sure that you explain to the reader which data were used in the analysis and how the different groups were compared to one another.

2. Check that you have detailed section headers

Section headers are important because they help the reader know exactly what will be described within a specific section. In the methods and results sections, you want to make sure that a reader will be able to quickly identify what each specific subsection is about. Headers are also important because they break up the text into smaller chunks, which makes it easier for a reader to digest the information. Sometimes journals will allow you to use section headers in the discussion, which can also help with segmenting material for your reader. Chapter 7 provides more details about section headers.

3. Check that you have included introductory and concluding sentences

Introductory and concluding statements within paragraphs are important for a coherent flow from section to section. Most manuscript and reports tell a "story" in some respect, in that one experiment result led to the development and execution of a second experiment. To that end, introductory and concluding statements help guide the reader. For example, avoid starting paragraphs with "Figure 1 showed..." Instead, introduce the topic and question being addressed before going into the details of your results. You can also provide a short description of the method that was used as well before going into the details of the results. Chapter 7 has additional information on introductory and conclusion statements.

4. Ensure that figures and tables could stand alone

Some readers will jump straight to looking at the tables and figures. For this reason, it is important that everything is defined within either the figure or table itself or within the legend. This includes making sure that you have redefined abbreviations in the legend, as well as provided clear keys, headers, and labels. Additionally, within a figure legend you can also add one or two sentences about the methodologies for each set of experimental data that you show.

5. Emphasize the takeaway message in the discussion

Some readers may jump to the discussion if they want to immediately see the main point of the paper. Therefore, in the last paragraph of the discussion, make sure you provide your readers with at least a one- to three-sentence overview of the entire study. Make sure these last concluding statements are general enough that someone who may not have necessarily read the entire paper can still understand the stated conclusion. See Chapter 10 for more details about writing the discussion section.

A last read-through checklist

Here are some general questions to ask yourself as you are editing each of the sections that you have written. Many of these ideas have been mentioned in previous chapters.

Introduction

- *Did you describe the problem you were trying to address?*
- *Did you describe why the problem has not yet been addressed yet (i.e., shortcomings of other studies)?*
- *Did you provide a hypothesis to your readers?*
- *Did you provide enough contextual information for the reader to understand your work?*

Methods

- *Did you describe all your experiments and analyses?*
- *If a reader were to look at data in a table or figure, would they be able to locate the associated methods in the methods section?*
- *Did you define your statistical analysis?*
- *Did you provide manufacturer names?*

- *Did you provide descriptions of regulatory approvals?*

Results

- *Do all your results make sense?*
- *Are all figures and tables referenced it the text?*
- *Are the data presented in figures and tables consistent with what is in the text?*
- *Are there too many numbers reported in the text of the results section?*
- *Are there any sentences that just list data? If yes, consider rewording these sentences and referencing a table or figure instead.*
- *Do you present your experiments in a logical order?*

Discussion

- *Did you recap your results?*
- *Did you discuss how your results compare to others in the field?*
- *Did you discuss how your results advance the field?*
- *Did you discuss limitations of your study?*
- *Did you discuss what some potential next steps are?*
- *Did you conclude the paper with an overall recap of the main finding?*

14
SELF-EDITING: LANGUAGE, STYLE, AND CLARITY

This chapter will provide some general language rules you should follow when preparing a manuscript, thesis, dissertation, or any other general scientific report. Many of the topics presented in this chapter are those that I have encountered in my editing of scientific and clinical manuscripts.

It is better to check for grammar issues in later drafts rather than the first draft. As you are doing your initial writing, you will make better use of your time if you focus on the content, rather than structure, that you are writing about. When you are ready to edit for structure and clarity, follow the guidelines in this chapter to ensure that you have a clear and well-written document.

Use objective language

Subjective language is language that has an emotion or personal attitude affiliated with the statement. In a conversation with other researchers, you may say something along the lines of "drug A is *better* than drug B" and others may understand what you mean by "better". However, this type of language should generally be avoided in scientific writing. Instead, you should describe exactly what is meant by the word "better". Does drug A result in reduced tumor growth? Does drug A prolong survival? Be specific. For example, you might describe the result as "mice treated with Drug A survived longer than mice treated with Drug B". Below is another example:

Subjective: Patients that did not receive treatment had worse outcomes than those who did.

Objective: Patients that received treatment had an increase in overall survival by 12 months compared to those without treatment (36 vs 24 months, P = .02).

Be specific and guide your reader

Guiding your reader is particularly important

when you are writing your results section. If you are describing data in your results, be sure to tell your reader which question you are addressing and what data you are referring to. See below for an example of vague and specific language.

Vague: Use of a therapeutic increased T cell localization in the tumor.

Specific: There was a higher proportion of CD8+ T cells within CD45+ tumor infiltrating lymphocytes in mice treated with Therapeutic A (35% CD8+) and Therapeutic B (38% CD8+) compared to the control group (12% CD8+; $P < .001$ for both comparisons).

Keep sentences focused

Depending on the topic you are discussing, you may find yourself with long and winding sentences. While the content may make sense to you, long sentences may be difficult for a reader to comprehend, especially if they are being introduced to the topic for the first time. Try to keep a sentence confined to a single topic, and break sentences up if necessary. Confining sentences to a single topic can also help you avoid grammatical errors such as run-on sentences or comma splices, either of which may occur when you try to string too many thoughts together.

In addition, be aware of how your paragraphs are structured. Paragraphs that are between four and 10 sentences long are easier for readers to digest than paragraphs that are 30 sentences.

Avoid sentence data dumps

In the process of assembling your results section, it may be difficult to figure out exactly how you want to present your data. It is important that you write your data and results in a way that will be easily digestible to readers. Otherwise, you may end up with sentences that are a bit difficult to read because they have too much data in them (in other words, a "data dump"); consider the following example:

Model A had a sensitivity, specificity, accuracy, positive predictive value, and negative predictive value of 84%, 79%, 80%, 26%, and 97% for lung lesion characterization. Model B had sensitivity, specificity, accuracy, positive predictive value, and negative predictive value of 74%, 94%, 92%, 49%, and 98%.

Presenting results in this manner can make it hard for a reader to compare the different models. In addition, in a list of written numbers, the reported data are not next to their associated metric, which again makes the data difficult to digest. As the above is written, there are also no statistical analyses presented nor is there any indication of which model had a statistically higher performance. Another, more effective, way to present this data could be like this:

Model B had higher performance than Model A in specificty (94% vs 79%; P = .02), accuracy (92% vs 80%; P < .001), and positive predictive value (49% vs 26%; P = .03). Model A had a higher sensitivity than Model B (84% vs 74%; P = .01) and we found no evidence of a different in the negative predictive values (97% vs 98%; P = .30).

Definitions and abbreviations

You may find that you used a lot of abbreviations in your protocols and results as you ran your experiments. While there are many abbreviations that may be common in a field (for example, CAR [chimeric antigen receptor] in immuno-oncology, or PET [positron emission tomography] in radiology), there are also many abbreviations that are very niche or specific to your study. For example, EPR could mean "enhanced permeability and retention" in the field of nanotherapeutics, or it could mean "electron paramagnetic resonance" in imaging. There might even be a paper about nanoparticles used in imaging where both abbreviations of EPR would technically be correct. In these scenarios, spelling out the abbreviation is best to increase reader clarity.

Of course, these examples are not inclusive. As

a general rule of thumb, try to only use between five to 10 abbreviations in your paper. This will make the paper easier to read. There are, however, times when more than 10 abbreviations may be appropriate. For example, a paper about cardiac imaging may have many abbreviations that are common in the field for cardiac anatomy (left ventricle [LV], right ventricle [LV], aorta [Ao], ascending aorta [AAo], right coronary artery [RCA], mitral valve [MV], etc.), and use of all these abbreviations may be appropriate. You will need to assess which abbreviations may be less familiar to your reader—these are the abbreviations you should write out.

If a long and unwieldy phrase is used more than two to four times in a paper, it is best to abbreviate it, even if it is a less well-known abbreviation. For example, if you use the disease name "cerebral autosomal dominant arteriopathy with subcortical infarcts and leukoencephalopathy" a few times in a manuscript or report, it is probably best to just abbreviate it as CADASIL after the first mention.

Avoid vague words

Does your manuscript have a lot of sentences that start with "this", "that", "these", "those", or "it", without a noun or a subject afterwards? In general, it is best to avoid the use of "naked" demonstrative pronouns without an associated subject.

Many times, these pronouns are being used to refer to the subject of a previous sentence. But again, we do not want to make our reader have to figure out which specific subject we are talking about. If you use these pronouns make sure that you also provide the subject within the same sentence for enhanced clarity. See a few examples below:

No subject included: In recent years there has been a trend of an increased number of residents interested in machine learning. To support this, the institution has implemented a new program around these topics.

Subject included: To support this <u>trend,</u> the institution has implemented a new program around <u>machine learning applications.</u>

No subject included: Overfitting can occur when a model is trained with a relatively small training dataset. Solutions to this include training with a larger dataset or using data augmentation.

Subject included: Solutions to <u>overfitting</u> include training with a larger dataset or using data augmentation.

Words and phrases to consider substituting

The next couple of pages contain a list of common words and phrases that you should consider searching for and replacing prior to submitting your manuscript or report. Using proper wording and removing any bulk or unnecessary wording will make your paper easier to read.

- *Statistically significant*
  - If a significant P value (i.e., > .05) and the phrase "statistically significant" are both in the same sentence, remove the phrase "statistically significant". As the P value already shows the significance, providing the extra explanation is not necessary. Removing this phrasing can help to reduce word count as well.

- *Better, best, worst,* and *worse*
  - Be specific. What is meant by better or worse? Was some metric higher or lower than another? Instead of using these general terms, state the specific trend with the associated data and statistical statements.

- *Novel, superior, first of its kind, state-of-the-art*
  - For the most part, any original research article is just that—original, new, novel, etc. For that reason, descriptions such as "novel" or "first of its kind" should generally not be included within your manuscript. It is assumed that your research is new.

- *Trending towards significance* or *almost significant*
  - You may have come across a paper that used the word "trending" to indicate that two results were *almost* significantly different, with an associated *P* value above the given alpha level (usually 0.05). However, you should avoid the use of "trending towards significance" or "almost significant" in your writing. In scientific literature, a relationship is generally regarded as significant or not, there are no "almost" significant trends.
  - If an alpha level is set at 0.05, this means that $P \leq .05$ is significant and $P > .05$ is not. To put this in context, saying that a $P = .08$ is trending towards significance is like saying $P = .02$ is trending towards non-significance.
  - Or you can simply state the P value with the associated data and leave the interpretation up to your future readers.
  - See the end of Chapter 9 for a list of statistical analysis resources.

Further Reading:

- Knight, K. L. & Ingersoll, C. D. Optimizing scholarly communication: 30 tips for writing clearly. J. Athl. Train. 31, 209-213 (1996).

15
WRITING A LETTER TO THE EDITOR

Once you have finished writing your manuscript and have confirmed the content with all coauthors, you are at a point where you can start to write your letter to the editor. This letter allows you to tell the journal editorial board about your work and why it should be published in their journal. Along with the title and abstract, this letter will be one of the first pieces of text the editor will read about your manuscript.

Sometimes the project PI may write this letter, but I highly encourage that graduate students, postdocs, and other laboratory researchers volunteer to write this letter—or at least provide a first draft to the PI to edit. Writing a letter to the editor is an excellent way to expand your writing skills. The purpose of this letter is to tell the editor exactly why you think your paper is a good fit for their journal and what your research can contribute to the current field.

To write this letter, you will need to know the correct article type, as well as the name of the editor. Some journals may have a few different manuscript types, but they generally fall within three categories: Original Research, Brief Report, and Review. Journals could have up to 10 variants, so you need to determine which category your manuscript falls under. The name of the editor can typically be found on the journal website.

Now that you have the editor's name and manuscript type, you have enough information to write the first paragraph of the letter. It is okay if this first paragraph is only a sentence or two. The first paragraph of the letter will look something like this:

Dear Dr. Editor,

We would greatly appreciate your consideration of "My Manuscript about Science" as an original research article in Super Science Journal.

In general, you should keep the letter to two to four paragraphs—you do not want it to be too long.

This letter should be written in a way that will allow the editor to quicky understand what the study is about and how it contributes to the field. The general outline for three paragraphs should be: (1) introduction and background, (2) what you did, and (3) why it is important.

Within this letter, you have the opportunity to recommend reviewers. Some journals may have a place on their submission site for these names, but they can also be added within the letter. Additionally, some journals will allow you to list individuals that you do not want to review your manuscript (non-recommended reviewers). You can put your list of recommended and non-recommended reviewers at the very end of your letter.

It is important to be aware of the people that you put down on these lists. If you work closely with or have multiple collaborators with Dr. X, it is a best practice to not suggest Dr. X as a reviewer because (1) they will likely be biased in assessing your manuscript, (2) if the peer review process is blinded, there is a good chance that Dr. X will figure out who the research group is, and (3) the manuscript may not be as critically assessed as it would be from a different reviewer. Some journals are moving away from recommended reviewers due to the potentially biased views of these selected reviewers.

Below are two examples of actual letters that I submitted in the past for manuscripts that were later accepted:

Dear Editor,

We would greatly appreciate your consideration of the manuscript, "Novel engraftment and T cell differentiation of human hematopoietic cells in Art$^{-/-}$ IL2RG$^{-/-}$ SCID pigs", for publication as a Research Article in Frontiers in Immunology.

Our group at Iowa State University discovered and described the first pigs with severe combined immunodeficiency (SCID) in 2012 that we are now developing for a variety of different biomedical models. We have already been successful in using this model for ovarian cancer, skin grafts, and stem cell research in collaborations across the United States. One of the major steps in developing this model is to immunologically humanize these animals through the transplantation of human hematopoietic stem cells. Humanization of the SCID pig would expand their utilization into fields related to cancer immunotherapies, HIV, and others that are reliant on intact human immune cells to study.

In this manuscript we describe our approach to improve the pig SCID model through gene editing and provide evidence of the first successful humanization attempt in this second generation SCID pig model. Our group at ISU is on the forefront of SCID pig model generation. We were the first to design and utilize biocontainment facilities for SCID pigs, as well as provide a detailed description of how to rear these animals. We have also published detailed procedures for in utero injection protocols, which we utilized within this submitted manuscript.

We believe that this will be a foundational manuscript as the newly emerging field of SCID pig research continues to develop. There are many researchers around the U.S. that are interested in the SCID pig model, and we believe this manuscript will emphasize new ways SCID pigs can be used in biomedical research. This project highlights the One Health Initiative as there are scientific researchers, veterinarians, medical professionals, and transplantation experts that have all been involved in this project.

All authors have seen and approved the submission of this manuscript. Thank you for your consideration.

Below is another example:

Dear Editors,

We would greatly appreciate your consideration of the manuscript, "Development of severe combined immunodeficient (SCID) pig models for translational cancer modeling: future insights on how humanized

SCID pigs can improve preclinical cancer research", for publication as a Hypothesis and Theory article in Frontiers in Oncology - Molecular and Cellular Oncology as part of the "Humanized" Large Animal Cancer Models: Accelerating Time and Effectiveness of Clinical Trials Research Topic.

Within the last decade, pigs with severe combined immunodeficiency (SCID) have been developed to serve as a preclinical animal model that better represents human anatomy and physiology than commonly used rodent models. We describe in this review many research applications that SCID pigs could be used for in the field of oncology. SCID pigs accept xenografts of human cancer cell lines. Such xenotransplantation in the pig model can be used to study tumor physiology, targeting imaging techniques, and drug testing. Additionally, major efforts are being made to generate an immunologically humanized SCID pig through human hematopoietic stem cell transplantation. We describe potential methods that could be utilized to humanize pigs based on previous research in immunocompetent pigs and sheep, as well as rodent models.

We believe that the dissemination of applications of SCID pigs in cancer research will increase awareness to cancer researchers that this model exists. We hope that this Hypothesis and Theory article will inspire new ideas of utilizing SCID pigs in different facets of cancer research.

All authors have seen and approved the submission of this manuscript.

Further Reading:

- Peh, W. C. & Ng, K. H. Writing a letter to the Editor. Singapore Med. J. 51, 532-535 (2010).

16
SUBMITTING YOUR MANUSCRIPT

Confirm your final draft with your coauthors

There are a few important steps that you should take before submitting your manuscript. Throughout the writing process you likely corresponded numerous times with your coauthors, but it is critical that all authors see the final version of the manuscript before it is submitted. Because their names are on the paper, those individuals are associated with the data and with the claims made in the paper as well. In other words, your coauthors are also responsible for the content.

In the past when I have finished a manuscript, I have sent a version out to all coauthors to confirm and to provide any last edits before submission. If you have a lot of coauthors, you may be handling quite a few final edits, so you can ask coauthors to edit on the same document (using an application such as Google Docs or Dropbox) if that is possible. This confirmation step is important in the submission process; often journals will not only have you confirm the author list, but they will also have you confirm that all coauthors have seen and approved the submitted version of the manuscript.

Double-check journal formatting

Another important step before submitting your manuscript is to double-check the journal format. Different journals will have different word count restrictions, figure and table limits, abstract structures, and other specifications. It is possible that you may have changed your mind at some point during the writing process as to which journal to submit to. Therefore, before you submit, you will want to double-check the guidelines of the journal.

A journal will likely not outright reject a paper if it is in the wrong format, but you may be notified to change the formatting before the manuscript can be considered by the editorial board. Having to reformat your document can prolong the review process. Therefore, it is better to check the formatting prior to submission. I have had papers returned to me because they were over by 400 words; in those cases, I had to edit and resubmit the paper. Remember to read the submission guidelines thoroughly.

Proofread your work

It may sound obvious, but it is essential that you read your manuscript all the way through prior to submitting to check for typos, formatting inconsistencies, missing punctuation, and accurate spelling. All of these checks fall under proofreading. Reading your manuscript out loud can be helpful in catching any sentences that may be missing words or punctuation.

The following are some other points to consider as you proofread:

- Confirm that figures and tables are cited in the text
- Double-check that all numerical data reported in the text from the figures and tables matches the values that are within the figures and tables
- If you performed an animal or human study, make sure that all patient and animal numbers are clearly defined for your different experiments
- Ensure that you have statements of ethical approval (Institutional Animal Care and Use Committee for animal studies and Institutional Review Board for human studies)

- Make sure your results as well as your materials and methods sections have sub-headers (if the journal allows). These sub-headers make it much easier for readers to see what the different sections are about.

17
THE EDITORIAL PROCESS AFTER SUBMISSION

Now that you have submitted your manuscript, you may wonder what happens next. Some journal websites will have an interface allowing you to see the different stage that your manuscript is in. The process may vary depending on the journal, but most follow the same general workflow. This chapter provides an overview of the process.

The types of editors

There are a few different types of editors involved in the editorial process that I would like to introduce before diving into the details of the editorial process:

Editor in Chief (EIC): This is the lead editor of the journal. The EIC assesses all manuscripts that are submitted to a journal and makes decisions regarding general journal policies. The EIC guides conversations with other editors regarding the manuscript and helps to make the final decision on a manuscript (along with the deputy editors, described in the next section).

Deputy Editors (DEs): This is a team of editors that also assess the submitted manuscript. Typically, each deputy editor has their own subspecialty field, so incoming manuscripts are divvied up based upon subject matter. For example, in a radiology journal focused on cancer, there may be one DE that is an expert in breast imaging, another DE that focuses on neurological tumors, etc.

Associate Editors (AEs): Associate editors help the editorial boards by contributing ideas for journal initiatives, as well as providing their expertise for certain manuscripts. If a paper is submitted to a journal that is very specialized, an editorial board may reach out to an AE for their opinion and assessment of the work.

Scientific Editors: These editors help to fine-tune manuscript readability by rearranging content and rewriting or editing throughout to improve clarity. Not all journals have a scientific editor, so you may not encounter this type of editing.

Manuscript Editors: Manuscript editors modify the language that is used in the journal articles and ensure style consistency among manuscripts of the same journal.

The workflow of manuscript submissions

1. Formatting check

Once a manuscript is submitted, the editorial office will take a first look at the manuscript. The editorial office makes sure that all components of the manuscript have been properly submitted. They are looking for general formatting of the manuscript and will make sure that the necessary sections are present (abstract, introduction, materials/methods, results, discussion, and references); additionally, they will check word counts. Their job is to make sure that the manuscript is complete before they send it off to the DEs and EIC. If there are sections of the manuscript that do not conform to the journal's guidelines, the manuscript will be sent back to the authors to fix. If the manuscript is sent back to you at this point, it typically does not have any bearing on whether it will be accepted or rejected—the EIC likely has not seen it yet.

2. DE assignment and decision

Once the manuscript has gone through the initial check, it will be assigned to a DE that has a matching background or expertise in the topic of the manuscript. They will look at the content of the manuscript and decide if it is a good fit for the journal. At this point, the DE is assessing the study design and significance of the results within the context of the research field. Once they make their decision to reject or continue pursuing the manuscript, their decision is then forwarded to the EIC.

3. EIC decision

All manuscripts are routed through the EIC. In the next stage of the process, the EIC will receive an initial decision from the DEs. If the manuscript does not seem to be a good fit for the journal, the manuscript will be rejected at this point. If the editors decide to continue with the manuscript, then it will go out for peer review.

4. Peer review

Peer review is when your manuscript is assessed by other experts in the field for scientific rigor, accuracy, and its potential impact. There are typically staff in editorial offices that can search for reviewers based on their different expertise areas. Journals usually send manuscripts out to two to four peer reviewers. Reviewers will assess the manuscript, provide comments for the authors, and then determine what the next steps should be for the manuscript. They can suggest acceptance, acceptance with minor revisions, acceptance with major revisions, or rejection. Depending on the journal, there may be other alternative revision statuses as well.

The peer review process takes time. Some reviewers may decline to peer review, which can add to the total time it takes for your manuscript to be reviewed. Reviewers may decline for simple reasons such as not having enough time during a given period. If reviewers decline to review, then other potential reviewers will be contacted.

5. Compile reviews, DE and EIC discussion, and final decision

Once the peer reviews have been received, the DEs and EIC assess the responses. During the peer review process, reviewers will have made different suggestions ranging from simple grammar changes to requesting additional discussion; in some cases, they will even request further experimentation.

Most manuscripts will go through at least one round of reviews and revisions. If a manuscript received a decision of "minor revision", there is a chance that the manuscript will not be sent back out to reviewers a second time. In these cases, the DE and EIC evaluate the revised manuscript. If a major revision was required, though, the manuscript will likely go back out to reviewers for another round of assessment.

In the end, the deciding factor of manuscript acceptance comes down to three main factors:

1. The author's ability to appropriately address reviewer and editor concerns
2. The accuracy and impact of the science and methodology in the field
3. If the study is aligned with the journal's publishing goals

6. Scientific editing and copyediting

Scientific and general manuscript editing can either be done before or after the manuscript has been accepted. This is typically a free service for all manuscripts that are accepted by the journal, as the journal wants to ensure that all content is thoroughly checked before publishing. If the journal you submitted to has scientific and manuscript editors, you will likely receive a version of your manuscript with queries related to different editing topics (grammar, content, figure and/or table organization, abstract, general methods, etc.). Depending on the journal, you will have anywhere from 48 hours (general manuscript editing) to a few weeks (scientific editing) to respond to the comments.

7. Publication of your manuscript

Once your manuscript has been accepted and has undergone editing, then it is nearly ready to be published. Editorial offices have production staff that transform your word-processed document into a PDF that follows the style of the journal. This is typically the last document you will be able to review before the manuscript is published. Some journals may publish weekly, monthly, or even bimonthly. How often a journal publishes may be variable; this information can usually be found on the journal's website.

Further Reading:

- Zelmer, J. Demystifying the Publication Process. Healthc. Policy 12, 6-9 (2017).
- Zucker, I. H. Demystifying the publishing process: a primer for early career investigators. Am. J. Physiol. Circ. Physiol. 309, H529-H531 (2015).

18
RESPONDING TO REVIEWERS

The basics

After a manuscript has been submitted, the typical review period can last anywhere from one to five months (or even longer), so prepare yourself to wait. Usually, manuscripts are reviewed by two to four different people. After waiting so long for feedback, it can be overwhelming to read and figure out how to respond to reviewers. In addition, it may be possible that your manuscript will go through more than one review and revision cycle.

I have typically handled reviewer comments in three "stages". During the first stage, I open the reviews and often find myself quite frustrated because my work (which I spent so much time on) has been nitpicked. After this initial glance, I usually do not look at the reviews again for a few days. During the second stage, I reread the peer reviews and come to realize that they are not that bad, which is when I can transition to preparing myself to work on the revision. The third stage is when I can begin to address reviewer comments and start revising the manuscript.

There are typically three categories of review and/or edit requests that you will see from reviewers:

1. Grammar text edits
2. Content text edits
3. Addition of new data (i.e. additional experiments) and/or explanations

Categories 1 and 2 are the easiest to handle; if the requests fall into category 3, though, you need to decide what steps to take next. That is to say, you can either perform the experiments or not. Sometimes a category 3 review can turn into category 2 edit; this is discussed in more detail later in this chapter.

Formatting your responses

Before diving into the details of how to respond to each of the categories, I will address how to effectively answer reviewer comments. You want the edits you make to be clear and easy for your reviewers (as well as the journal editors) to identify. There are two main ways that this can be done. One approach is to use track changes within your word-processor program to specifically mark where the edit has been made.

Usually, peer reviewers will numerically list their comments within their review. It is best to reference the specific comment number in your revised manuscript. For example, if you are addressing comment 5 from reviewer 1, you should use the notation of "R1-5" within your revised manuscript where the associated edit or change was made. This is a nice and easy way for the reviewer to go back to the original document and see exactly where changes have been made.

An additional approach would be to list line numbers within the document as you are writing your responses. Usually, a word-processor program will have the option to list line numbers within the document (some journals require this for submission anyway), so as you are answering the reviewers in your response sheet, you can call out the specific lines in which you made the edit. It can be frustrating for the reviewer if they cannot easily see your changes—a happy reviewer is better than an upset reviewer!

Strategies for starting your revision

One strategy that I use when responding to reviewers is to type out my response to each reviewer comment *before* starting to edit my manuscript. The responses I write are not well written at first; instead, they act as a draft and help to provide me with a plan for addressing specific changes that I know I need to make. These responses can even be written as bullet points initially and then can be polished later when you are closer to resubmitting.

I use these written responses and bullet points as a framework for my future editing. I find that I have a clearer head when I can sit down and just focus on answering the reviewer comments and questions than if I had to sit down and start editing right away.

Another strategy you could use is to go through your document and highlight problem areas within your document. This again will help you lay a foundation for determining what areas of your manuscript you need to give attention to as you edit.

Category 1 reviewer comments

Category 1 comments are the easiest to address. Sometimes reviewers will identify typos or inadvertently missed information (abbreviations, company names for products, etc.), and all you need to do is add in the relevant information. Sometimes the reviewers will ask for clarification on certain topics throughout the paper. These are also typically very quick and minor changes. Here are some examples of category 1 comments:

- References are out of order
- Typos in different areas of the paper
- Company names for products, reagents, or equipment are needed
- Protein and gene text styling is incorrect
- Too many run-on or short sentences
- Incorrect Table of Figure number callouts in the text

Category 2 reviewer comments

Category 2 comments are those that require additional information to be added to the manuscript. These comments are usually focused on addressing limitations in the study or providing additional information and context for specific results of your study. Typically, you will need to provide additional descriptions about these topics within your manuscript.

Listed on the next page are some examples of category 2 comments:

- Clarification of control or experimental groups in text is needed
- Expanding on the limitations of your study
- Providing additional details on the broader impacts or implications of your work
- Describing necessary background information required to understand your study

Category 3 reviewer comments

The last type of comment is one that suggests that you should add new data. Journals will typically try to avoid asking you to perform new large experiments, as this is not a trivial task; however, sometimes these comments may still make it into the final review. If you are asked to add new data, there are typically three options you have for moving forward: (1) perform the experiment, (2) address it as a category 2, or (3) withdraw the paper and resubmit it elsewhere. The decision you make will vary on a case-by-case basis. Sometimes the reviewer asks for an experiment similar to one that you have already done. In this case, you can treat it as a category 2 and provide additional information and content about the experiment and overall concept. If the experiment is relatively easy and can be performed in a timely manner, then doing the experiment may not be an issue.

Sometimes the experiment that is proposed by

the reviewer is not one that your research group is interested in doing; you may not have the staff or time to conduct it, or you may feel strongly that the paper does not need the experiment. In these cases, it may be best to withdraw the manuscript, edit, and resubmit it elsewhere.

Disagreeing with reviewers

While reviewers are experts in their subject areas, they are still human and may make mistakes. Additionally, reviewers may have left comments that you do not agree with. If you come across a mistake or a comment that upsets you, be mindful of the tone of your response to the reviewer. It can sometimes be easy to let your feelings get in the way, and your responses to reviewers may come across as rude or potentially unprofessional in tone.

Remember that you are talking to a person; the reviewer also has feelings. It does not do you any good to have a negative tone in your responses; the reviewers are less likely to see your edits in a positive light if there is negative language in your responses. For that reason, it is good practice to reread your responses and then let a collaborator read them as well. Remember to also thank the reviewers for their insights and improvements to your manuscript. While responding to reviewers can be difficult, manuscripts do improve as a result of the reviewers' feedback.

That said, I know that there can be instances when reviewers have a less-than-pleasant tone in their comments to authors. Typically, editors will remove these unnecessary comments from the letter that they send to you. However, if you come across a reviewer response that seems unnecessary or uses derogatory language, communicate with the Editor in Chief about these issues. The comments that you receive should all be objective in nature—as should your responses.

Submitting your revision and what to expect next

For resubmission of your manuscript, you should typically have three documents prepared:

1. A document with your responses to the peer reviewer comments
2. An edited manuscript with tracked changes and comments
3. A clean manuscript in a publish-ready state with all changes accepted

Since you are sending two copies of the manuscript, be sure to emphasize to the reviewers and editors exactly which document they should be looking at when they read your responses. For example, you may have decided to reference line numbers on the clean copy of the manuscript rather than on the marked-up document. You want to make sure that the reviewers know where to look when they are reading over your responses.

When the reviewers initially submitted their responses to the journal, they may have opted to see the manuscript once more (i.e. the reviewer wants to see your future revision after you have responded to their comments) before making an accept or reject decision, or they may have decided that the decision could be made by the editorial board of the journal. There is really no easy way for the author to know this, as it is not readily available on the online journal dashboards. However, you can email the editor for more information about the status as well as the process that your manuscript will go through upon resubmission. Depending on what the reviewers have opted for, it may affect the time it takes for you to hear back after resubmission. Remember that long wait periods do not equate to rejections.

Some examples of reviewer comments and responses

Here are a few fictitious examples of reviewer comments and my responses based on some of my previously published works. In general, the nature of reviewer comments will vary depending on the specific topic, but these examples show some general topics that reviewers may comment about.

1. Sometimes reviewers ask for something that is

already in the paper. This likely means that you did not clearly describe what you did (or the reviewer read too fast and missed it). In the example below, the reviewer asked for an experiment that I had already performed and presented in the original submission of the manuscript.

REVIEWER: Fig. 5 shows that pig macrophages do not phagocytose human cells. A key question not addressed within your study is if cross-reaction between human Protein A and pig Protein B is responsible for the resistance of human cells to phagocytosis by pig macrophages? This can be easily addressed by comparing phagocytosis of human cells by pig macrophages with or without the blocking antibody.

RESPONSE: We performed this experiment and documented the results in the original Figure 6. We have clarified and updated Figure 6 in the revised version of this manuscript (detailed in lines 300-325 of the clean document).

2. Below is an example of a category 3 comment. Instead of performing another experiment, we added a section to the discussion describing the limitation.

REVIEWER: The human myeloid populations generated in the humanized mouse are not fully characterized. Some other markers (preferable by PCR) are required to define the development of those populations.

RESPONSE: Thank you. We have added section "Further characterization of differentiated human immune cells" to the discussion to address the limitations of this study and to emphasize the importance of in-depth characterization of human immune cells in future humanization studies. Found on lines 600-620 in the CLEANCOPY document.

3. Reviewers will also find your typos and let you know.

REVIEWER: There is a minor typo in the abstract sentence: "Additionally, we and successfully..." Please correct.

RESPONSE: We have fixed the typo in the abstract.

4. Be aware of your abbreviations. Using too many can be distracting and make your paper hard to read. Reviewers will point this out as well.

REVIEWER: Some abbreviations should be indicated (e.g., SRY gene, CTNNB1 gene, HEPES buffer).

RESPONSE: These abbreviations have been expanded and defined throughout the text.

5. Make sure you mention the companies and recipes for the solutions you used in your study. These details are extremely important for reproducibility.

REVIEWER: Methods are well executed/composed but there is no consistent information given on companies for some material/machines (sometimes no information, sometimes yes). Please indicate. Please specify composition of the stem cell expansion medium and indicate company.

RESPONSE: We have added company names throughout. We have added the recipe for the media within the methods section, lines 500-512 in CLEANCOPY document.

Further Reading:

- Wong, G. L.-H. Tips for Responding to Reviewers' Comments-from an Editor's or Reviewer's Points of View. Gut and liver vol. 13 7-10 (2019).
- Jain, P. & Patwari, A. K. Responding to Reviewers Comments. Indian Pediatr. 53, 1093-1095 (2016).

19
ACCEPTANCE AND LAST STEPS FOR PUBLISHING

You have finally gotten your acceptance letter, so what comes next?

First, celebrate! It is a great achievement to get a manuscript accepted. However, there are still a few things that will happen as your paper moves from "accepted" to "published". As mentioned earlier (in Chapter 17), the journal now needs to transform your document to match their journal's style.

Receiving proofs before publishing

Depending on the journal, the actual publishing

of your manuscript can take one week or up to four or five months, so it may be a while before you hear back from journal staff. If you want to know the timeline, there are usually journal staff that you can contact who will give you this information.

When your publication date is close, you will receive a proof of your manuscript in the style and format of the journal. Some journals have manuscript editors that will copyedit your manuscript for language and grammar. The proof that you receive will typically have these copyedits included with queries from the copyeditors. Most journals typically allot 48 hours for you to review your proofed manuscript. You will also want to take a read through it one more time to ensure that any typos have been fixed before final publication—you will not get another opportunity to make changes. Have other coauthors of the manuscript read at the proof too.

Typically, there will be specific queries from the manuscript editors for you to answer. Such queries may have statements such as the following:

- "This sentence has been edited for clarity, please confirm that the edited sentence retains your intended meaning."
- "Please provide an expansion of this abbreviation"

- "Please confirm all author names are correct and that authors are listed in the appropriate order"

These queries are generally easy to answer. Journals will usually request that comments responding to these queries are placed directly within the proof.

Paying the publication fee

Depending on the journal that you submitted your work to, you may have to pay a publishing fee. Some journals do not have fees, while others charge in the range of $50 to $500 USD; or you may have to pay an article processing fee, which could range from $2,000 to $8,000 USD for open access articles. (There may be journals that charge more or less than these ranges). These fees help to pay for the cost of publishing the article (from editing to production of the manuscript for online publishing and website uploading, etc.). As the project you were working on was most likely funded by a grant, these publication fees will come out of that pool of money. Some institutions have transformative agreements with publishers such that the institution's library may pay the article processing fee for open access publishing, and the researchers from those institutions can then publish for "free". Talk with your PI or laboratory lead about publication fees if your journal requires them.

Other small tasks

Depending on the nature of the journal or your article, there may be a few more things that you need to do upon the acceptance of your paper:

- If your paper describes the use of a large dataset, there may be a requirement that this dataset be made publicly available, in which case you would need to find a website to host this material.

- If you used any figures from a previously published paper (regardless of whether it was your own or someone else's), you need to make sure that you have permission from the other journal to reprint it. Figures from journals are usually copywritten and require permission and fees for reprinting. If an article was published open access under a Creative Commons license, you may only need to cite the material within your manuscript. Double-check the type of article you are reusing content from whenever you decide to use this content in your own projects.

20
HANDLING REJECTION

It is likely that you have seen or heard of someone having a manuscript rejected from a journal. Having a rejection come your way can be stressful and feel disheartening for a paper that you have spent so much time on. However, it is important to take the rejection as an opportunity to improve the manuscript. The sooner you can come to terms with the rejection, the sooner you can start to identify objective ways in which you can edit and work on the manuscript for resubmission to another journal.

Within this chapter, I will discuss reasons that your manuscript may be rejected. Sometimes when you receive a rejection, there may not be a reason listed on the letter you receive from the journal. You may receive an outright rejection (i.e., you need to submit to a different journal), or you might receive a decision of "rejection with option for a major revision", in which case you need to decide if you want to revise and resubmit or do a de novo submission to another journal.

Regardless of which kind of rejection you receive, you and your collaborators need to think critically and objectively about your manuscript to figure out why it may have been rejected.

Understanding the reason for the rejection will help you determine if you need to:

- Submit to another journal
- Conduct some more experiments before resubmission
- Rethink the idea you are trying to publish and potentially rework the project prior to submitting again.

The next few sections outline the major reasons that a manuscript may be rejected.

The subject matter does not match the journal

One reason that you may receive a rejection is that the content of your article is not a good fit for the journal. While it may have seemed to you that your content *did* match the journal (after all, that is why you submitted there), the editors of the journal may think that the readership of their journal will not be as interested in the content of your article. Keep in mind that this is not an attack on your study. This type of rejection can be a particular challenge for researchers that are working in emerging fields because these fields may be a niche that does not yet have a relevant journal.

For example, I worked with swine biomedical models in graduate school. At the time I was working on the research and getting ready to publish, there were not a lot of researchers (i.e., editorial board members and peer reviewers) that were familiar with swine models. Most biomedical research is performed on mice, and publishing studies using mice is the norm for most journals. Thus, our group had some difficulties in identifying journals to submit to.

If you find that you need to reassess your journal choice, search a few journal databases with key words that are related to your manuscript. Determine which journals publish similar content and consider submitting your manuscript to these journals instead.

Insufficient data or analysis

Another reason that your manuscript might be

rejected is if the editorial board feels that you have insufficient data for the conclusions that are being drawn. Depending on the field that you are in, it may be difficult to have a high sample number for the system you are studying; however, numbers are important in your conclusions. If there are a small number of patients in a study to test drug efficacy, there might not be enough statistical power to come to a concrete conclusion.

If you find yourself in this scenario, you have a couple of options. First, you may need to go back to the drawing board and plan another round of experiments. While this may not sound fun, it may be necessary in order for your study to have more statistical power and/or evidence for the conclusions being drawn. A second option may be to reframe your work. Depending on what field you are in and the journal you are submitting to, there may be options to submit work as brief reports or technical developments. In these types of manuscripts, usually small sample sizes are more acceptable because a new technology or methodology is being developed. Again, perform a literature search to see if you can identify any manuscripts like yours that have been published to help you determine which is the best option for you.

Subject content or methodology is not timely

Sometimes a lab may be in a situation in which

they have had data available for a few years, but no one was able to put it together in a manuscript in a timely manner (lack of staff, time, etc.). When there are finally individuals available who can put the manuscript together, the original data may no longer be relevant. For example, imagine a study about a drug that was the standard of care for breast cancer 10 years ago, but which is no longer a standard of care drug. Such a study may no longer be clinically relevant, so a journal will reject it because it is not timely. To avoid a rejection for this reason, assess the data that you have and consider when it was collected and if it is still relevant.

Writing is unclear or confusing

If a manuscript is presented in such a way that a reviewer or editorial board cannot figure out what the main point of the paper is, the paper may be rejected. If someone reads through a paper and is confused by it, they may be left wondering why they should publish it. In these instances, it is best to have other individuals read your paper so that they can guide you through which parts of the paper could be clarified. Many journals also have a list of editing companies that you can reach out to for external editing services.

Further Reading:

- Venkatesh, S., Maymone, M. B. & Vashi, N. A. Peer reviews: the dreaded rejection. Dermatol. Online J. 24, (2018).
- Ali, J. Manuscript rejection: causes and remedies. J. Young Pharm. 2, 3-6 (2010).

21
OTHER RESOURCES

<u>Guidelines and Editing Resources</u>
Here are a few guidelines and editing resources that you can use to double-check your manuscript prior to submission. Some journals will require that you provide these guidelines and indicate that you have included all relevant content described in the guidelines along with your submission. Even if a journal does not require that you use these guidelines, they can be helpful to look through as you are finalizing your manuscript.

<u>ARRIVE guidelines</u>
Animal Research: Reporting of In Vivo Experiments
https://arriveguidelines.org/

CoBRA guidelines
Citation of BioResources in journal Articles
BMC Med. 2015. 13:33. PMID: 25855867

CONSORT guidelines
Consolidated Standards of Reporting Trials
BMJ. 2010: 340: c332. PMID: 20332509

GPP2 guidelines
Good Publication Practice
BMJ. 2009. 339: b4330. PMID: 26259067

ICMJE
International Committee of Medical Journal Editors
http://www.icmje.org/

MIBBI guidelines
Minimum Information for Biological and Biomedical Investigations
https://fairsharing.org/collection/MIBBI
This website has almost 40 different checklists ranging different types of biological studies (T cell assays, phylogenetic analysis, biofilm experiments, peptide array experiments, etc.).

PRISMA-DTA guidelines
Preferred Reporting Items for Systematic Reviews-Diagnostic Test Accuracy
JAMA. 2019. 322(20):2026. PMID: 29362800

STARD guidelines
Standards for Reporting Diagnostic Accuracy
Radiology. 2015:151516. PMID: 26509226

STROBE guidelines
Strengthening the Reporting of Observational Studies in Epidemiology
Int J Surg. 2014. 12(12): 1500-24. PMID: 25046751

See NIH Research Reporting Guidelines and Initiatives for more resources
Website:
https://www.nlm.nih.gov/services/research_report_guide.html

22
WHAT DID I MISS?

Thanks for reading this guide!

As we all know, there are many steps involved in the writing and editing process. While this guide does provide an overview on many of these steps, there are likely some topics that you wanted to read about that were not within this guide.

If you have a topic that you would have liked to have seen within this guide, let me know!

Visit the site below and leave a message about the topic that I forgot to include!

adelineboettcherbioblog.com/i-forgot

ABOUT THE AUTHOR

Adeline Boettcher, PhD is a professional scientific writer and editor. She earned her PhD in Molecular, Cellular, and Developmental Biology from Iowa State University in 2019. She has an interest in preclinical and translational biomedical research. She owns Alpha Beta Scientific Communications LLC, a scientific writing, editing, and consulting company.

-Notes-

-Notes-

-Notes-

-Notes-

-Notes-

-Notes-

-Notes-

-Notes-

-Notes-

-Notes-

-Notes-

-Notes-